Selected Titles in This Series

(*Continued in the back of this publication*)

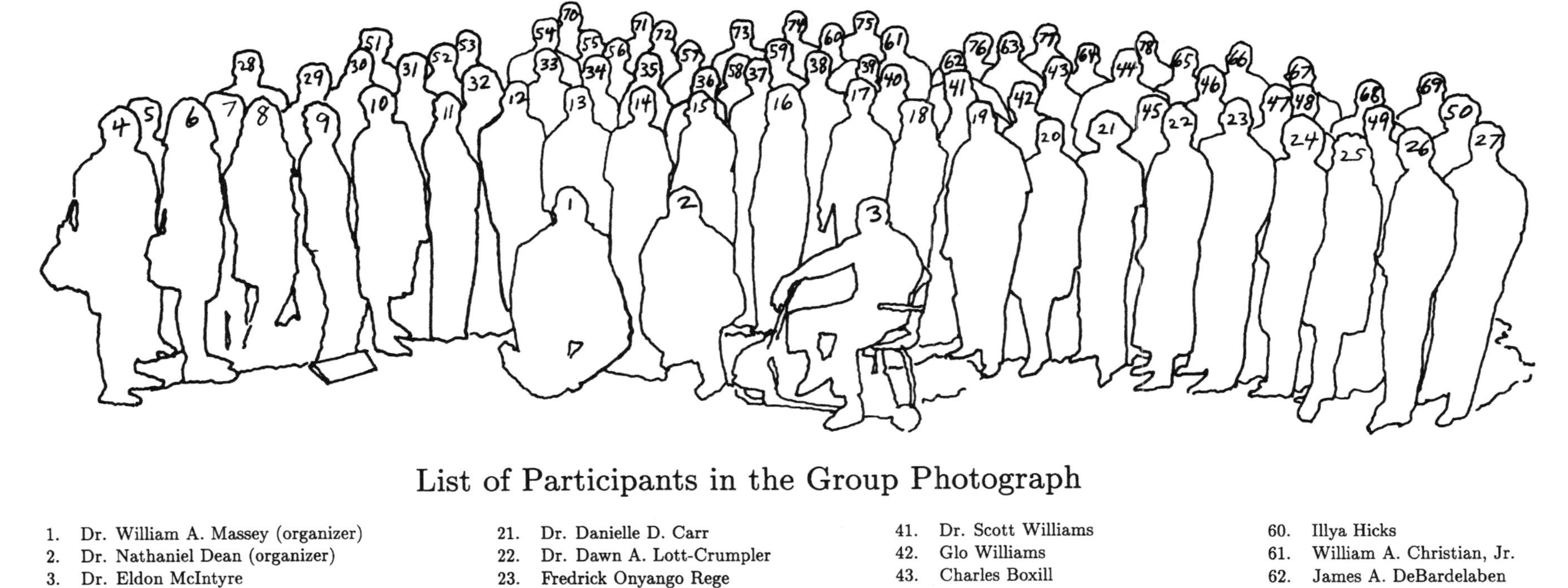

List of Participants in the Group Photograph

1. Dr. William A. Massey (organizer)
2. Dr. Nathaniel Dean (organizer)
3. Dr. Eldon McIntyre
4. Tamika Thompson
5. Katrina K. Ashford
6. Sophia Reid
7. Djuana Lea
8. Cassandra McZeal
9. Louise Brown
10. Dr. Stella Ashford
11. Carolyn Coleman
12. Kossi Delali Edoh
13. Alain Togbe
14. Dr. Oneaka Mack-Humphrey
15. Gregory L. Allen
16. Diana Dismus-Campbell
17. Robert C. Stolz
18. Dr. Camille A. McKayle
19. Dr. Earl R. Barnes
21. Dr. Danielle D. Carr
22. Dr. Dawn A. Lott-Crumpler
23. Fredrick Onyango Rege
24. Atiya N. Hoye
25. Tasha R. Inniss
26. Jason M. Lewis
27. Kori Needham
28. Alfred Noel
29. Dr. Floyd Williams
30. Daryl R. Brydie
31. Dr. Don Hill
32. Angela E. Grant
33. Rod Moten
34. Dr. Johnny L. Houston
35. Dr. Curtis Clark, Jr.
36. Dr. Fern Hunt
37. Dr. Melvin R. Currie
38. Sirbrittie Grant
39. Martin Khumbah
41. Dr. Scott Williams
42. Glo Williams
43. Charles Boxill
44. Dr. Leon C. Woodson
45. Dr. Sylvester Reese
46. Kimani Stancil
47. Harold Young
48. Dr. Charles Thompson
49. Robert McDonald
50. Dr. Carolyn Mahoney
51. Asamoah Nkwanta
52. Lemuel R. Riggins
53. Brad Gray
54. Dr. Isom Herron
55. Michael O. Keeve
56. Dr. Donald R. King
57. Michael Gatlin
58. Nonetta M. Pierre
59. Dr. Lee Lorch
60. Illya Hicks
61. William A. Christian, Jr.
62. James A. DeBardelaben
63. Otis B. Jennings
64. Gene Jarrett
65. Craig J. Sutton
66. Dion Stevens
67. Tyrone E. McKoy, Jr.
68. Dr. Debra Curtis
69. Idris Stovall
70. John Sims
71. Dr. Nathaniel Whittaker
72. Dr. Arthur Grainger
73. Dr. Jonathan David Farley
74. Dr. Donald F. St. Mary
75. Dr. James A. Donaldson
76. Cyril Coumarbatch
77. Mark Lewis
78. Jimmie L. Davis, Jr.

DIMACS

Series in Discrete Mathematics
and Theoretical Computer Science

Volume 34

African Americans in Mathematics

DIMACS Workshop
June 26–28, 1996

Nathaniel Dean
Editor

NSF Science and Technology Center
in Discrete Mathematics and Theoretical Computer Science
A consortium of Rutgers University, Princeton University,
AT&T Labs, Bell Labs, and Bellcore

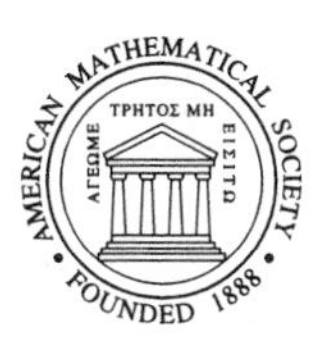

American Mathematical Society

This DIMACS volume includes papers by invited speakers and poster presenters at the Second Conference for African-American Researchers in the Mathematical Sciences, held at DIMACS June 26–28, 1996. It also includes papers on issues related to African-American involvement in the mathematical sciences.

Supported in part by the Sloan Foundation and AT&T Labs.

1991 *Mathematics Subject Classification.* Primary 00B15, 00B25, 01A80.

Library of Congress Cataloging-in-Publication Data

Conference for African-American Researchers in the Mathematical Sciences (2nd : 1996 : Center for Discrete Mathematics and Theoretical Computer Science)

African Americans in mathematics : Second Conference for African-American Researchers in the Mathematical Sciences, June 26–28, 1996 / Nathaniel Dean, editor.

p. cm. — (DIMACS series in discrete mathematics and theoretical computer science, ISSN 1052-1798 ; v. 34)

Held at the Center for Discrete Mathematics and Theoretical Computer Science (DIMACS) at Rutgers University in Piscataway, N.J.

"NSF Science and Technology Center in Discrete Mathematics and Theoretical Computer Science, a consortium of Rutgers University, Princeton University, AT&T Labs, Bell Labs, and Bellcore."

Includes bibliographical references.

ISBN 0-8218-0678-5 (alk. paper)

1. Mathematics—Congresses. 2. Afro-American mathematicians—Congresses. I. Dean, Nathaniel, 1956– . II. NSF Science and Technology Center in Discrete Mathematics and Theoretical Computer Science. III. Title. IV. Series.

QA1.C623 1996

510—DC21 97-21748

CIP

---

10 9 8 7 6 5 4 3 2 1 02 01 00 99 98 97

Contents

Part III. Historical Articles

Foreword

The "Second Conference for African-American Researchers in the Mathematical Sciences" was held at DIMACS, the Institute for Advanced Study, and the research facilities of Bell Labs and AT&T Labs in Murray Hill in June 1996. DIMACS expresses its thanks to Nate Dean and William Massey for their tireless efforts to organize this program and produce this volume. DIMACS was honored to be a partner in this conference dedicated to showcasing and advancing research by African-American mathematicians.

DIMACS gratefully acknowledges the generous support that makes these programs possible. The National Science Foundation, through its Science and Technology Center program; the New Jersey Commission on Science and Technology; DIMACS partners at Rutgers, Princeton, AT&T Labs, Bell Labs, and Bellcore generously support all DIMACS programs. We would like to express our thanks to the Institute for Advanced Study, the Sloan Foundation, AT&T Labs, and Bell Labs for additional support and use of facilities that contributed successfully to making this program one of national significance.

Fred S. Roberts
Director

Bernard Chazelle
Co-Director for Princeton

Stephen R. Mahaney
Associate Director

Preface

The Second Conference for African-American Researchers in the Mathematical Sciences was held for three days at the Center for Discrete Mathematics and Theoretical Computer Science (DIMACS) at Rutgers University in Piscataway, New Jersey, June 26-28, 1996. It was organized by Nathaniel Dean and William A. Massey, both of Bell Laboratories, the research division of Lucent Technologies. The main goal of the conference was to highlight current research by African-American researchers and graduate students in mathematics, to strengthen the mathematical sciences by encouraging the increased participation of African-American and underrepresented groups, to facilitate working relationships between them, and to help cultivate their careers.

We had over 100 researchers and graduate students in attendance who were exposed to a variety of technical and cultural events. Participants were introduced to some of the major research centers in New Jersey: DIMACS at Rutgers University in Piscataway, the Institute for Advanced Study (IAS) in Princeton, as well as Bell Laboratories and AT&T Labs who were both located in Murray Hill. Visiting all these research institutions was a first for most of the participants. There were twelve one-hour invited technical talks given by researchers spanning a variety of mathematical and scientific disciplines. At IAS we held group discussions, led by Fern Hunt (NIST) and Camille McKayle (Lafayette College) that focused on issues surrounding minority participation in mathematics, such as: The Career Life Cycle of an African-American Mathematician; Jobs of the Present, Jobs of the Future; The Public Image of Mathematics and Mathematicians in the African-American Community; and Affirmative Action. At Murray Hill, a select group of 17 graduate students presented their current research during the poster session where they interacted in smaller groups with conference attendees as well as researchers both from Bell Labs and AT&T Labs. This volume includes papers by the invited speakers and poster presenters as well as papers on issues related to African-American involvement in the mathematical sciences.

We wish to thank the staff at DIMACS for helping to organize and host this event. We thank DIMACS, the Sloan Foundation, and AT&T Labs for providing funds, and we thank DIMACS, Bell Labs, and IAS for the use of their facilities. We would also like to thank the participants of the conference, the authors, the anonymous referees, and Christine M. Thivierge of AMS for helping with this event and the preparation of this volume.

Nathaniel Dean & William A. Massey
March 1997

Part I

Invited Research Talks

DIMACS Series in Discrete Mathematics
and Theoretical Computer Science
Volume **34**, 1997

Chain Decomposition Theorems for Ordered Sets
and Other Musings

Jonathan David Farley

This paper is dedicated to the memory of Prof. Garrett Birkhoff

ABSTRACT. A brief introduction to the theory of ordered sets and lattice theory is given. To illustrate proof techniques in the theory of ordered sets, a generalization of a conjecture of Daykin and Daykin, concerning the structure of posets that can be partitioned into chains in a "strong" way, is proved. The result is motivated by a conjecture of Graham's concerning probability correlation inequalities for linear extensions of finite posets.

1. Introduction

Order has played a rôle in human civilization for as long as the North Star has hung above our heads. The theory of ordered sets, however, is a relatively new discipline.

Lattice theory and the theory of ordered sets are exciting areas with a number of surprising connections with other branches of mathematics, including algebraic topology, differential equations, group theory, commutative algebra, graph theory, logic, and universal algebra. Both fields have many important applications, for example, to scheduling problems, the semantics of programming languages, the logic of quantum mechanics, mathematical morphology and image analysis, circuit design, and cryptography **[2]**, **[10]**, **[14]**, **[21]**.

Below we present a sampling of theorems — some famous, some not — dealing with decompositions of ordered sets into chains. We will expand upon our own original work (presented in §7) in a future note.

1991 *Mathematics Subject Classification.* Primary 06A07, 05A18; Secondary 06-02, 06-06, 05-02, 05-06.

The author would like to thank Dr. William Massey and Dr. Nathaniel Dean for inviting him to speak at the Second Conference for African-American Researchers in the Mathematical Sciences. The author would also like to thank Prof. Bill Sands for permitting him to publish his theorem. This paper is based on a lecture given at the Institute for Advanced Study in Princeton, New Jersey, on June 27, 1996.

2. Basic terminology, notation, and examples

A *partially ordered set*, or *poset*, is a set with a binary relation $\leqslant$ that is reflexive, transitive, and antisymmetric. Elements a, $b \in P$ are *comparable*, denoted $a \sim b$, if $a \leqslant b$ or $a \geqslant b$; otherwise they are *incomparable*, denoted $a \parallel b$. An element a is a *lower cover* of b if $a < b$ and $a < c \leqslant b$ implies $b = c$ for all $c \in P$; in this case, b is an *upper cover* of a, which we denote by $a \prec b$. The *Hasse diagram* of a finite poset is a graph in which each vertex represents an element, and an edge drawn upward from a to b means that $a \prec b$. Hence $a \leqslant b$ if and only if one can go from a to b by tracing the edges upward.

A *chain* is a subset C of P such that $a \sim b$ for all a, $b \in C$. A chain $c_1 < c_2 < \cdots < c_n$ is *saturated* if $c_1 \prec c_2 \prec \cdots \prec c_n$. The *length* of a finite chain is $\#C - 1$ (where $\#C$ is the cardinality of C). The *height* $\operatorname{ht} x$ of an element x is the supremum of the lengths of the finite chains with greatest element x. The height $\operatorname{ht} P$ of a poset is the supremum of the heights of its elements. A finite poset is *ranked* if, for all $x \in P$, every chain with greatest element x is contained in a chain of length $\operatorname{ht} x$ with greatest element x. A saturated chain $c_1 \prec c_2 \prec \cdots \prec c_n$ in a ranked poset is *symmetric* if $\operatorname{ht} c_1 = \operatorname{ht} P - \operatorname{ht} c_n$.

An *antichain* is a subset A of P such that $a \parallel b$ for distinct a, $b \in A$. The *width* of a poset is the supremum of the cardinalities of its antichains.

For all $p \in P$, let $\uparrow p := \{q \in P \mid p \leqslant q\}$ and $\downarrow p := \{q \in P \mid p \geqslant q\}$. An element p is *maximal* if $\uparrow p = \{p\}$; let $\operatorname{Max} P$ denote the set of maximal elements. An *up-set* of P is a subset U such that, for all $u \in U$, $\uparrow u \subseteq U$.

The *disjoint sum* of two posets P and Q is the poset with underlying set $P \cup Q$ and the inherited order on P and Q, but no comparabilities between elements of P and elements of Q. The *ordinal sum* of P and Q is the poset with underlying set $P \cup Q$ and the inherited order on P and Q, but with $p < q$ for all $p \in P$ and $q \in Q$. It is denoted $P \oplus Q$.

A *lattice* is a non-empty poset L such that, for all a, $b \in L$, the least upper bound of a and b exists, called the *join* of a and b (denoted $a \vee b$), and the greatest lower bound of a and b exists, called the *meet* of a and b (denoted $a \wedge b$). A *sublattice* is a non-empty subset closed under join and meet. A lattice is *distributive* if, for all a, b, $c \in L$, $a \wedge (b \vee c) = (a \wedge b) \vee (a \wedge c)$ [equivalently, for all a, b, $c \in L$, $a \vee (b \wedge c) = (a \vee b) \wedge (a \vee c)$].

An example of a chain is the real line with the usual ordering. The power set of a set is a distributive lattice; here join corresponds to set union and meet to set intersection. The collection of all subsets with 42 elements is an antichain. The collection of subsets of $\{x, y, z\}$ containing either $\{x, y\}$ or $\{z\}$ is an up-set of the power set of $\{x, y, z\}$.

The left and right sides of Figure 1 are the Hasse diagrams of the four-element fence P and the four-element crown Q, respectively. Their disjoint sum and ordinal sum are shown in Figures 1 and 2.

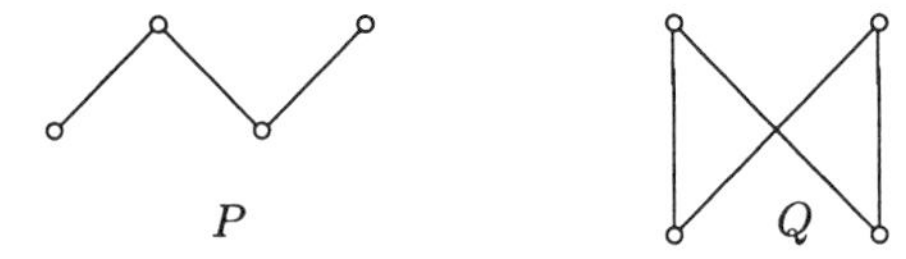

Figure 1. The disjoint sum of P and Q.

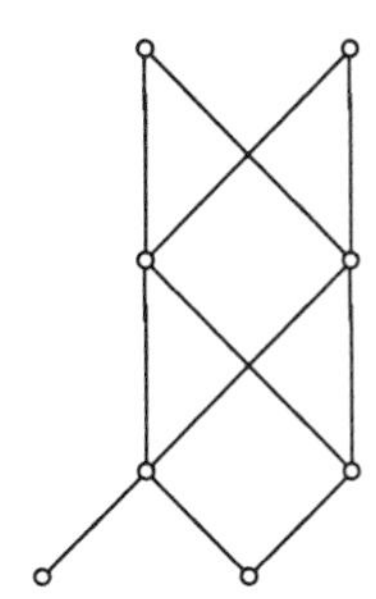

Figure 2. The ordinal sum $P \oplus Q$.

Another example of a distributive lattice is the set of natural numbers ordered by divisibility ($a \leqslant b$ if a divides b). In this lattice, $a \vee b$ is the least common multiple of a and b, and $a \wedge b$ is the greatest common divisor of a and b.

The lattice of equivalence classes of propositions in propositional logic is a distributive lattice when ordered by implication (the class $[p]$ of the proposition p is less than or equal to $[q]$ if p implies q). In this lattice, $[p] \vee [q] = [p \text{ or } q]$ and $[p] \wedge [q] = [p \text{ and } q]$.

The lattice of subspaces of a vector space is *not* distributive if the dimension of the space is at least 2. The five-element lattices M_3 and N_5 of Figure 3 are also non-distributive.

Figure 3. The lattices M_3 and N_5.

Figure 4 shows the Hasse diagram of a poset that is not a lattice.

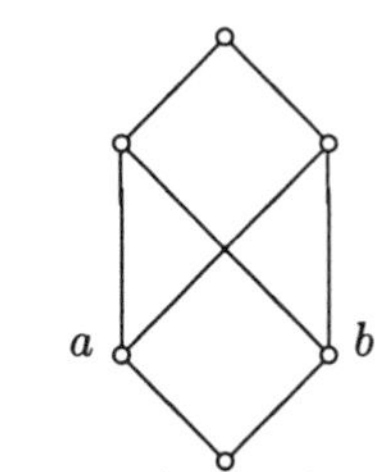

Figure 4. A non-lattice.

The elements a and b have common upper bounds, but not a *least* upper bound, so the poset is not a lattice.

Basic references on lattice theory and the theory of ordered sets are [**1**] and [**4**].

3. Dilworth's theorem

Even mathematicians who know very little about the theory of ordered sets have heard of Dilworth's Theorem. (R. P. Dilworth actually proved many beautiful and important theorems in the theory of ordered sets, all of them called "Dilworth's Theorem." We, in particular, should note that Dilworth helped to develop mathematics programs on the Continent [**3**].)

Suppose one desires to partition a poset into chains, using the smallest number possible. (For instance, one would want such a partition for some scheduling applications.) If P has finite width w, then clearly we cannot do better than w chains. Dilworth's Theorem asserts that we can achieve this bound ([**8**], Theorem 1.1).

DILWORTH'S THEOREM. *Let P be a poset of finite width w. Then there exists a partition of P into w chains, and this is best possible.*

Note that we are not assuming P is finite; Dilworth's Theorem fails, however, for posets of infinite width, even if every antichain is finite [**17**].

For example, in Figure 5, the poset has width 4 ($\{a, b, e, f\}$ is an antichain), and can be partitioned into 4 chains (e.g., $\{0, a, d, 1\}$, $\{b\}$, $\{e\}$, $\{c, f\}$).

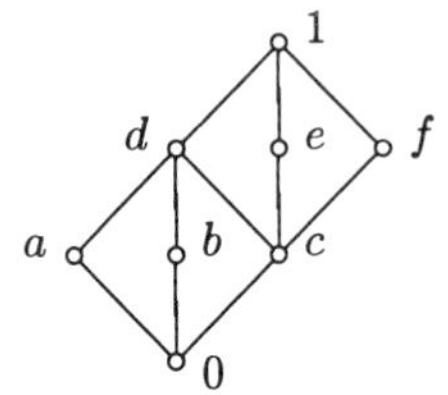

Figure 5. An illustration of Dilworth's Theorem.

For more on Dilworth's work, see [**9**].

4. Birkhoff's theorem

Garrett Birkhoff was one of the pioneers of lattice theory and universal algebra. Two of his many significant contributions involve distributive lattices.

Birkhoff characterized distributive lattices in terms of forbidden sublattices [analogous to Kuratowski's characterization of planar graphs in terms of forbidden subgraphs ([**11**], Theorem 6.2.1)].

THEOREM (Birkhoff–Dedekind). *A lattice is distributive if and only if neither M_3 nor N_5 is a sublattice.*

The theorem makes it easy to spot non-distributive lattices. For instance, the lattice of Figure 5 is not distributive. For a proof of the theorem, see [**4**], 6.10.

Now we know what lattices are *not* distributive. Which lattices *are* distributive?

Here is one way to construct distributive lattices: Take a finite poset, and order its up-sets by inclusion. This procedure yields a finite distributive lattice, in which join and meet are given by union and intersection, respectively.

For example, if Abubakari is the three-element fence (Figure 6), its lattice of up-sets is shown in Figure 7.

a
b c

Figure 6. The poset Abubakari.

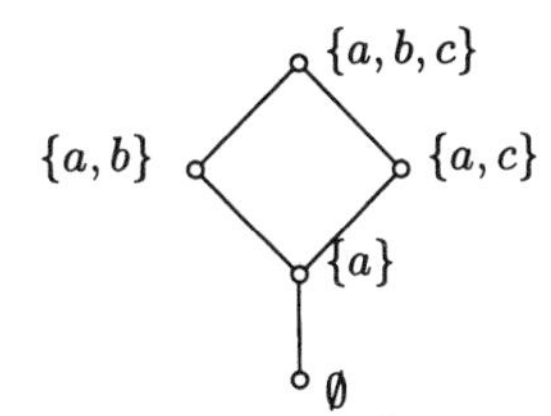

Figure 7. The distributive lattice of up-sets of Abubakari.

Another theorem of Birkhoff's asserts that *all* finite distributive lattices arise in this way ([**4**], 8.17). One consequence, which we shall need later, is that finite distributive lattices are ranked.

THEOREM (Birkhoff). *Every finite distributive lattice is the lattice of up-sets of a finite poset.*

There is an analogous theorem for infinite distributive lattices, but, instead of finite posets, one must use compact Hausdorff partially ordered spaces, called *Priestley spaces* [**18**].

5. Sands' matching theorem

Now we may prove an amusing theorem due to Bill Sands [**20**]:

THEOREM (Sands, 197*). *Every finite distributive lattice with an even number of elements can be partitioned into two-element chains.*

In the terminology of graph theory, the theorem asserts that the *comparability graph* of a finite distributive lattice with an even number of elements [in which (a, b) is an edge if $a \sim b$] has a perfect matching without loops ([**11**], §7.1).

The reader might like to attempt his or her own proof before proceeding.

PROOF. We prove the theorem by induction on the size of L. If $\#L$ is odd, and L has greatest element 1, we prove that $L \setminus \{1\}$ can be partitioned into two-element chains.

Represent the lattice as the lattice of up-sets of a finite poset P. Pick $p \in P$, and consider the the elements $a := \uparrow p$ and $b := P \setminus \downarrow p$ of L. Then L is partitioned into the finite distributive lattices $\uparrow a$ and $\downarrow b$. There are four cases to consider, depending on the parities of $\# \uparrow a$ and $\# \downarrow b$. □

The theorem is interesting because most chain-decomposition theorems apply either to arbitrary posets (e.g., Dilworth's Theorem) or only to power set lattices (e.g., [**16**]). Sands' Matching Theorem is obvious for power set lattices, but does not apply to all posets, not even all lattices. For example, M_4 (Figure 8) is non-distributive, by Birkhoff's theorem; it cannot be partitioned into two-element chains. (This example is due to Sands.)

6. The parable of the tennis players

Our main theorem is a generalization of a conjecture of Daykin and Daykin that deals with special sorts of chain-partitions. The motivation for the conjecture comes from a parable, due to Graham, Yao, and Yao ([**13**], §1):

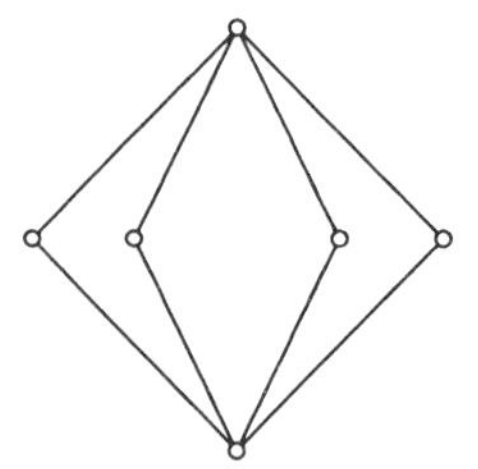

Figure 8. The lattice M_4 cannot be partitioned into two-element chains.

Imagine there are two teams of tennis players, A and B. The players of Team A are linearly ranked from best to worst, as are the players from Team B, but we know only in a few cases how individual players from Team A compare with individuals from Team B.

We may describe this set-up using a poset: The players are the elements, and a relation $p < q$ means that player p is worse than player q. Hence the subset corresponding to Team A is a chain, as is the subset corresponding to Team B (so that we have a poset of width at most 2). For an example, see Figure 9.

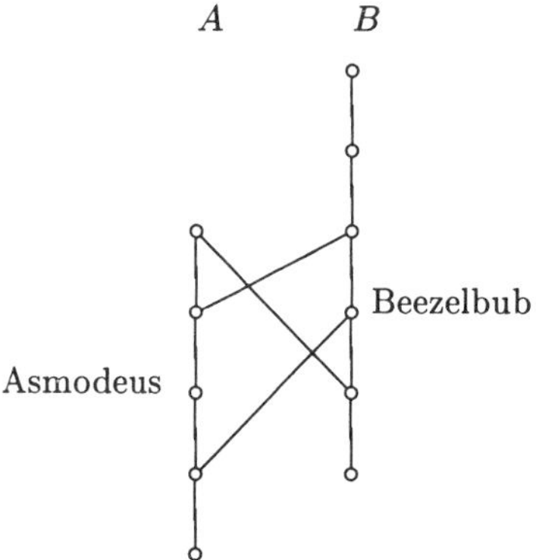

Figure 9. A tale of two tennis teams.

In Figure 9, we do not know whether Asmodeus (in Team A) is better or worse than Beezelbub (in Team B). We might ask for the *probability* that Asmodeus is worse than Beezelbub, Pr(Asmodeus<Beezelbub). We calculate this probability by looking at all the possible linear rankings of the players in both teams that are consistent with what we already know about which players are better than which.

Formally, we are looking at all the possible *linear extensions* of the poset, the bijective order-preserving maps from P into a chain of cardinality $\#P$. By counting the number of these in which (the image of) Asmodeus is below (the image of) Beezelbub, and dividing by the total number of linear extensions (assuming all are equally likely), we have Pr(Asmodeus<Beezelbub).

Now suppose we are given additional information. Namely, suppose we learn that certain players from Team A are worse than certain players from Team B. (Perhaps the two teams have just finished playing a tournament.) This information supports the idea that the players from Team A are worse than the players from Team B, making it more likely that Asmodeus is, in fact, worse than Beezelbub.

Formally, we would expect:

$$Pr(\text{Asmodeus}<\text{Beezelbub}) \leqslant Pr(\text{Asmodeus}<\text{Beezelbub} \mid a < b \;\&\cdots\&\; a' < b'),$$

the conditional probability that Asmodeus is worse than Beezelbub given that a is worse than b, a' worse than b', etc. (See Figure 10.)

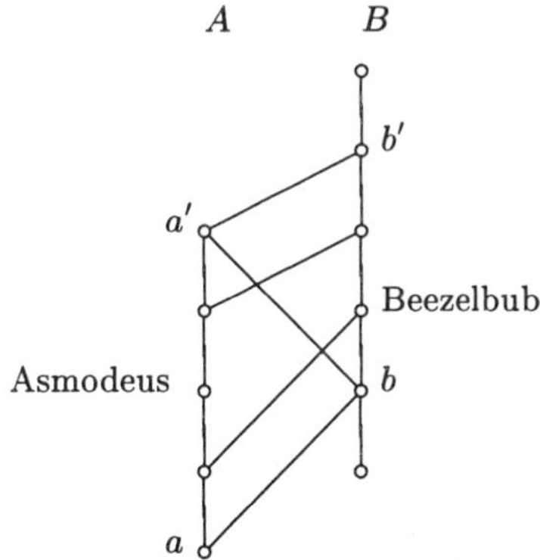

Figure 10. The poset of tennis players after a tournament.

We can formalize a more general situation. Let P be a poset partitioned into subsets A and B (not necessarily chains). Suppose that, whenever x and y are disjunctions of statements of the form

$$a < b \;\&\cdots\&\; a' < b'$$

we know that

$$Pr(x)Pr(y) \leqslant Pr(x\&y).$$

Then we say the partition has the *positive correlation property.*

Using a result of Daykin and Daykin ([**5**], Theorem 9.1), a question of Graham's ([**12**], pp. 232-233) becomes:

QUESTION (Graham). *Let $P = A \cup B$ be a partition of a finite poset such that, for all $a \in A$ and $b \in B$,*

$$a < b \textit{ implies } P = \uparrow a \cup \downarrow b$$

and

$$a > b \textit{ implies } P = \downarrow a \cup \uparrow b.$$

Does the partition have the positive correlation property?

7. A conjecture of Daykin and Daykin

In light of Graham's question, Daykin and Daykin ([**5**], §9) conjectured the following:

CONJECTURE (Daykin–Daykin). *Let P be a finite poset partitioned into chains T_1, T_2, and T_3 such that, if p and q are in different chains and $p < q$, then $P = \uparrow p \cup \downarrow q$.*

Then P is an ordinal sum $R_1 \oplus \cdots \oplus R_n$ such that, for $1 \leqslant i \leqslant n$, either

1. *R_i is disjoint from some T_j, $j \in \{1, 2, 3\}$, or*
2. *for all p, $q \in R_i$, if p and q are in different chains, then $p \parallel q$.*

With this conjecture, Daykin and Daykin asserted that certain correlation inequalities would follow.

Figure 11 shows an example of a poset satisfying the hypotheses of the conjecture. Let T_1 be the chain of all the elements on the left, T_2 all the elements on the right, and T_3 the remainder.

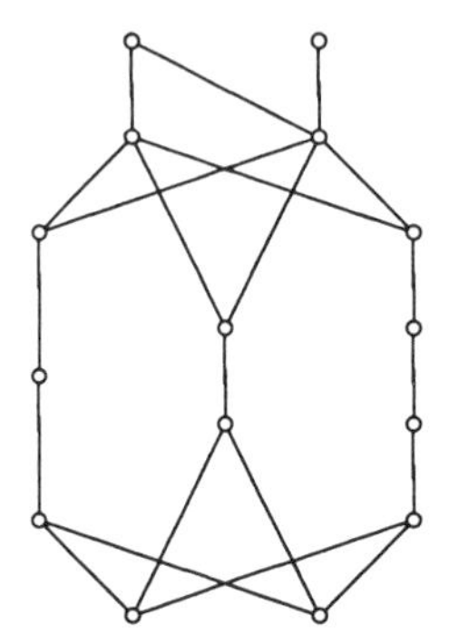

Figure 11. A poset satisfying the hypotheses of the Daykin–Daykin conjecture.

Note that the conclusion of the conjecture implies that P is the ordinal sum of width 2 posets and disjoint sums of chains.

Tseng and Horng [**23**] claim to have a proof, but their proof rests on a mistaken assertion. The author has proven a more general statement, with a much simpler proof, which applies to infinite posets; we shall write another note on this theorem. (Incidentally, although the fact was not publicized, it is reported in [**6**], p. 84 that J. M. Robson proved the Daykin–Daykin conjecture by an inductive argument.)

The results in the sequel are due to the author.

By $P(R_1, \dots, R_n; T_1, \dots, T_s)$, we mean:

1. P is a finite poset;
2. $P = R_1 \oplus \cdots \oplus R_n$;
3. $P = T_1 \cup \cdots \cup T_s$;
4. $\{ T_j \mid j = 1, \dots, s \}$ are disjoint chains;
5. for $i \in \{1, \dots, n\}$, either
   (6) R_i is disjoint from T_j for some $j \in \{1, \dots, s\}$ or
   (7) $p \parallel q$ for all $p \in R_i \cap T_j$ and $q \in R_i \cap T_k$ $(1 \leqslant j < k \leqslant s)$.

LEMMA. *Let $P'(R'_1, \dots, R'_m; T'_1, T_2, \dots, T_s)$ where $s \geqslant 3$. Suppose that $P = P' \cup \{x\}$ is a poset with a new element x, $\operatorname{ht} x = \operatorname{ht} P$ and for all $p \in P$ such that $\operatorname{ht} p = \operatorname{ht} P$,*

$$\#\{ q \in P \mid q \prec p \} \leqslant \#\{ q \in P \mid q \prec x \}.$$

Assume that $T_1 := T'_1 \cup \{x\}$ is a chain. Finally, assume that, for all $p \in T_j$, $q \in T_k$, $1 \leqslant j, k \leqslant s$, $j \neq k$, $p < q$ implies $P = {\uparrow} p \cup {\downarrow} q$.

Then $P(R_1, \dots, R_n; T_1, \dots, T_s)$ for some $R_1, \dots, R_n \subseteq P$.

PROOF. If $\operatorname{Max} P = \{x\}$, let $R_i := R'_i$ $(1 \leqslant i \leqslant m)$ and $R_{m+1} := \{x\}$. We may thus assume that $\operatorname{Max} P \neq \{x\}$.

We claim that $p \leqslant x$ for all $p \in R'_{m-1}$.

For if $p \nleqslant x$, where $p \in R'_{m-1} \cap T_k$, then ${\uparrow} p \subseteq T_k$; in particular, $R'_m \subseteq T_k$. If $C \subseteq P$ is a chain containing x, then $C \cup \{y\} \setminus \{x\}$ is a chain for all $y \in R'_m$, so that, for some $y \in R'_m$, $\operatorname{ht} y = \operatorname{ht} P$. Hence y has more lower covers than x, a contradiction.

If $R'_m \cup \{x\}$ is disjoint from some T_j, we finish by letting $R_i := R'_i$ $(1 \leqslant i < m)$ and $R_m := R'_m \cup \{x\}$.

We may therefore assume that $(R'_m \cup \{x\}) \cap T_j \neq \emptyset$ for $j = 1, \ldots, s$.

Case 1. For $1 \leqslant j \leqslant s$, $R'_m \cap T_j \neq \emptyset$.

Assume there exists $p \in T_2 \cap R'_m$ such that $p < x$. Then $R'_m \subseteq {\uparrow}p \cup {\downarrow}x$ and ${\uparrow}p \subseteq T_2 \cup \{x\}$, so that $(\operatorname{Max} P) \setminus \{x\} \subseteq T_2$. Hence $R'_m \cap T_3 \cap {\downarrow}x \neq \emptyset$, so that $(\operatorname{Max} P) \setminus \{x\} \subseteq T_3$. Hence $\operatorname{Max} P = \{x\}$.

Case 2. For some $j \in \{1, \ldots, s\}$, $R'_m \cap T_j = \emptyset$.

Clearly $j = 1$. If x has no lower covers in R'_m, then R'_m is an antichain and we are done. Otherwise, let $p \in R'_m$ be such that $p \prec x$ and $\{p, x\}$ belongs to a chain of length $\operatorname{ht} P$. Then for all $y \in \operatorname{Max} P$, $p \prec y$, and every lower cover of y is below x.

Now let $q \in R'_m$ be a lower cover of x. For all $y \in \operatorname{Max} P$, there exists $z \in P$ such that $q \prec z \leqslant y$ and $p < z$, so that $z = y$. Therefore we may let $R_i := R'_i$ $(1 \leqslant i < m)$, $R_m := R'_m \setminus \operatorname{Max} P$, and $R_{m+1} := \operatorname{Max} P$. □

THEOREM. *Let P be a finite poset and $T_1, \ldots, T_s$ disjoint chains covering P $(s \geqslant 3)$. Assume that, for all $p \in T_j$, $q \in T_k$ such that $p < q$, $P = {\uparrow}p \cup {\downarrow}q$ whenever $1 \leqslant j, k \leqslant s$ and $j \neq k$.*

Then there exist subsets $R_1, \ldots, R_n$ of P such that $P = R_1 \oplus \cdots \oplus R_n$ and, for $i = 1, \ldots, n$, either

1. *$R_i \cap T_j = \emptyset$ for some $j \in \{1, \ldots, s\}$, or*
2. *$p \parallel q$ whenever $p \in R_i \cap T_j$, $q \in R_i \cap T_k$, and $j \neq k$.*

PROOF. The theorem follows from the lemma by induction. □

The theorem will be generalized in a future note.

8. An open problem regarding chain decompositions of distributive lattices

Dilworth's Theorem asserts that, for every width w poset, there *exists* a partition into w chains. Both practical and philosophically-minded persons might prefer an explicit construction for such a minimal partition.

It is clear that any partition of a ranked poset into symmetric chains is a minimal chain-partition. Such a decomposition also tells us that the width of the poset is the number of elements of height $\lfloor \frac{1}{2} \operatorname{ht} P \rfloor$.

Examples of distributive lattices admitting such symmetric chain decompositions are finite power set lattices and lattices of divisors of positive integers [**7**].

Let $L(m, n)$ denote the poset of m-tuples $(x_1, \ldots, x_m)$ where $0 \leqslant x_1 \leqslant \cdots \leqslant x_m \leqslant n$, ordered as follows: $(x_1, \ldots, x_m) \leqslant (y_1, \ldots, y_m)$ if $x_i \leqslant y_i$ for $i = 1, \ldots, m$. This poset is a distributive lattice. Figure 12 shows $L(2, 3)$.

CONJECTURE (Stanley). *The lattice $L(m, n)$ admits a symmetric chain decomposition.*

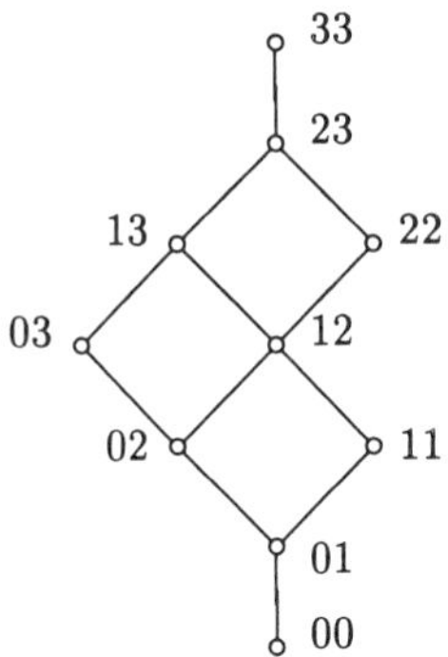

Figure 12. The lattice $L(2,3)$.

It is known that the conjecture holds for $L(3,n)$ and $L(4,n)$ [**15**], [**19**], [**24**]; it is also known that the width of $L(m,n)$ is indeed the number of elements of height $\lfloor \frac{1}{2} \operatorname{ht} L(m,n) \rfloor$, but the proof relies on heavy machinery from algebraic geometry ([**22**], §7). It would be interesting to have an elementary proof even of this fact.

9. Afterglow

"The Universe," said the philosopher with a glint in his eye, "is a big place." One might make a similar remark regarding the theory of ordered sets. What we have done is survey a tiny portion of the night sky. We hope that we have instilled in you, the reader, the desire to explore the constellations of lattice theory and the theory of ordered sets. Just perhaps, you may discover a star of your own.

References

1. Garrett Birkhoff, *Lattice Theory*, American Mathematical Society, Providence, Rhode Island, 1967.
2. Garrett Birkhoff and John von Neumann, *The logic of quantum mechanics*, Annals of Mathematics **37** (1936), 823–843.
3. Kenneth P. Bogart, *Obituary: R. P. Dilworth*, Order **12** (1995), 1–4.
4. B. A. Davey and H. A. Priestley, *Introduction to Lattices and Order*, Cambridge University Press, Cambridge, 1990.
5. David E. Daykin and Jacqueline W. Daykin, *Order preserving maps and linear extensions of a finite poset*, SIAM Journal on Algebraic and Discrete Methods **6** (1985), 738–748.
6. Jacqueline Wendy Daykin, *Monotonic Functions of Finite Posets*, Ph. D. Thesis, University of Warwick, 1984.
7. N. G. de Bruijn, Ca. van Ebbenhorst Tengbergen, and D. Kruyswijk, *On the set of divisors of a number*, Nieuw Archief voor Wiskunde **23** (1951), 191–193.
8. R. P. Dilworth, *A decomposition theorem for partially ordered sets*, Annals of Mathematics **51** (1950), 161–166.
9. Robert P. Dilworth, *The Dilworth Theorems: Selected Papers of Robert P. Dilworth*, Kenneth P. Bogart, Ralph Freese, and Joseph P. S. Kung (eds.), Birkhauser Boston, Inc., Boston, 1990.
10. B. Ganter and R. Wille, *Formale Begriffsanalyse: Mathematische Grundlagen*, Springer, 1996.
11. Ronald Gould, *Graph Theory*, The Benjamin/Cummings Publishing Company, Inc., Menlo Park, California, 1988.
12. R. L. Graham, *Linear extensions of partial orders and the FKG inequality*, Ordered Sets, Ivan Rival (ed.), D. Reidel Publishing Company, Dordrecht, Holland, 1982, pp. 213–236.
13. R. L. Graham, A. C. Yao, and F. F. Yao, *Some monotonicity properties of partial orders*, SIAM Journal on Algebraic and Discrete Methods **1** (1980), 251–258.

14. Henk J. A. M. Heijmans, *Morphological Image Operators*, Academic Press, Inc., Boston, 1994.
15. Bernt Lindström, *A partition of $L(3,n)$ into saturated symmetric chains*, European Journal of Combinatorics **1** (1980), 61–63.
16. Zbigniew Lonc, *Proof of a conjecture on partitions of a Boolean lattice*, Order **8** (1991), 17–27.
17. Micha A. Perles, *On Dilworth's theorem in the infinite case*, Israel Journal of Mathematics **1** (1963), 108–109.
18. H. A. Priestley, *Ordered topological spaces and the representation of distributive lattices*, Proceedings of the London Mathematical Society **24** (1972), 507–530.
19. W. Riess, *Zwei Optimierungsprobleme auf Ordnungen*, Arbeitsberichte des Instituts für Mathematische Maschinen und Datenverarbeitung (Informatik) **11** (1978).
20. Bill Sands, personal communication (1996).
21. Dana Scott, *Data types as lattices*, SIAM Journal on Computing **5** (1976), 522–587.
22. Richard P. Stanley, *Weyl groups, the hard Lefschetz theorem, and the Sperner property*, SIAM Journal on Algebraic and Discrete Methods **1** (1980), 168–184.
23. Shiojenn Tseng and Muh-Chyi Horng, *Partitions of a finite three-complete poset*, Discrete Mathematics **148** (1996), 205–216.
24. Douglas B. West, *A symmetric chain decomposition of $L(4,n)$*, European Journal of Combinatorics **1** (1980), 379–383.

Mathematical Sciences Research Institute, 1000 Centennial Drive, Berkeley, California 94720

Department of Mathematics, Vanderbilt University, Nashville, Tennessee 37240
E-mail address: farley@msri.org

DIMACS Series in Discrete Mathematics
and Theoretical Computer Science
Volume **34**, 1997

Unimodality and the Independent Set Numbers of Matroids

Carolyn R. Mahoney

ABSTRACT. In this paper we explore certain long-standing unimodal conjectures related to the sequence of independent set numbers of matroids and some progress toward their solution.

Matroid theory is an area of combinatorics that connects several branches of mathematics, for example, it encompasses certain areas of graph theory, linear algebra, lattice theory, group theory, ring theory, field theory, and combinatorial optimization. In this paper we introduce matroids, give examples of the relationship between matroids and matrices and matroids and graphs, and briefly explore some equivalent axiomatizations of matroids. We then consider certain long-standing unimodal conjectures related to the sequence of independent set numbers of matroids and some progress toward their solution.

1. Introduction

Recall [1] that a graph G is an ordered triple (V(G), E(G), φ_G) consisting of a nonempty set V(G) of vertices, as set E(G), disjoint from V(G), of edges, and an incidence function φ_G that associates with each edge of G an unordered pair of (possibly identical) vertices.

Example. Consider the graph pictured in Figure 1 with vertex set

$$V = \{v_1, v_2, v_3, v_4\}$$

and edge set

$$E = \{e_1, e_2, e_3, e_4, e_5, e_6, e_7, e_8\},$$

and φ_G defined by
$\varphi_G(e_1)=v_1v_1$, $\varphi_G(e_2)=v_1v_2$, $\varphi_G(e_3)=v_1v_4$, $\varphi_G(e_4)=v_1v_3$, $\varphi_G(e_5)=v_2v_4$, $\varphi_G(e_6)=v_3v_4$, $\varphi_G(e_7)= v_2v_3$, $\varphi_G(e_8)= v_2v_3$.

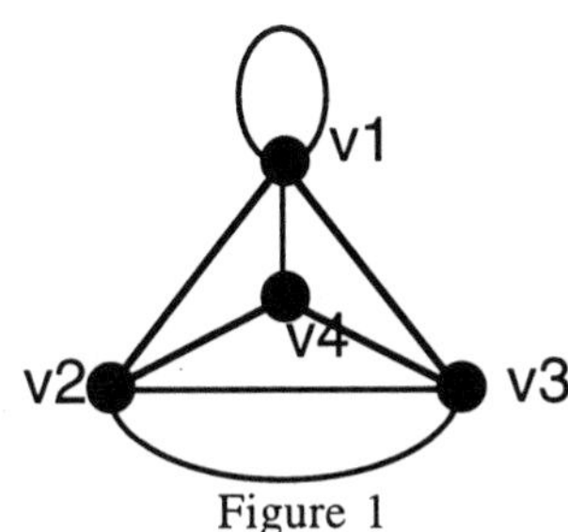

Figure 1

1991 *Mathematics Subject Classification.* Primary 05B35.

Edges like e_1 which begin and end at the same vertex are called loops; an edge with distinct endpoints is called a link. Pairs of link edges like e_7 and e_8 which have common endpoints are called parallel edges.

A path in a graph G is a non-null sequence $P = v_0e_1v_1e_2 \ldots e_kv_k$ whose terms are alternately vertices and edges such that for $1 \le i \le k$, the ends of e_i are v_{i-1} and v_i, and such that the vertices $v_0, v_1, \ldots, v_k$ are distinct. If $e = uv$ is an edge of G and $P = ue_1v_1\ldots e_kv$ is a uv path in G which avoids e, then the sequence $ue_1v_1\ldots e_kveu$ is called a *circuit* or a *cycle*. For example, $\{e_1\}$, $\{e_7, e_8\}$, $\{e_2,e_3, e_5\}$, and $\{e_2,e_3, e_6, e_8\}$are some of the circuits in the graph of Figure 1. Further basic information about graphs can be found in (1).

With a view toward matroids, we call the set of edges in a circuit a minimal dependent set. Notice, in the graph of Figure 1, that the empty set is not a minimal dependent set and that no minimal dependent set is contained in another minimal dependent set (hence the word minimal). It is also true that if two minimal dependent sets, say C_1 and C_2, have an edge e in common, then there exists another minimal dependent set, call it C_3, which is contained in the union of the first two, but avoids e (Take, for example,
$C_1 = \{e_2,e_3, e_5\}$, $C_2 = \{e_3,e_4, e_6\}$, and $e = e_3$ in the graph of Figure 1; we see that $\{e_2,e_4, e_5, e_6\}$satisfies the requirements for C_3)

In any graph, the set of edges that does not contain a minimal dependent set is called independent. It is easy to see that the empty set is independent, and that any subset of an independent set is independent. It is also true that if I_1 and I_2 are two independent sets such that $|I_1| < |I_2|$, then there is an element $e \in I_2 - I_1$ such that $I_1 \cup e$ is independent. In the graph above, we can take for example $I_1 = \{e_2, e_3\}$, $I_2 = \{e_3,e_5, e_6\}$, and see that $I_1 \cup e_6$ is independent.

Example. Let A be the matrix

$$\begin{matrix} 12345 \\ \begin{bmatrix} 10011 \\ 01001 \end{bmatrix} \end{matrix}$$

over the field R of Reals. Let E be the set of columns of A,

$$E = \{c_1 = \begin{pmatrix} 1 \\ 0 \end{pmatrix}, c_2 = \begin{pmatrix} 0 \\ 1 \end{pmatrix}, c_3 = \begin{pmatrix} 0 \\ 0 \end{pmatrix}, c_4 = \begin{pmatrix} 1 \\ 0 \end{pmatrix}, c_5 = \begin{pmatrix} 1 \\ 1 \end{pmatrix}\}.$$

In the case of matrices, linearly independent sets of columns will correspond to independent sets in the matroid, and minimal linearly dependent sets of columns will correspond to minimal dependent sets in the matroid. In the above example, then, the set of minimal dependent sets is Ç = $\{c_3\},\{c_1,c_4\},\{c_1,c_2,c_5\},\{c_2,c_4,c_5\}\}$. The set of independent sets is given by

$$I = \{\emptyset,\{c_1\},\{c_2\},\{c_4\},\{c_5\},\{c_1,c_2\},\{c_1,c_5\},\{c_2,c_4\},\{c_2,c_5\},\{c_4,c_5\}\}.$$

We can notice some properties of the collection of minimal dependent sets and the collection of independent sets in each of the examples described above. The properties we are interested in will give rise to two equivalent axiom systems for a matroid.

In the case of the minimal dependent sets, we observe that if Ç is the set of minimal dependent sets then

(C1) $\emptyset \notin$ Ç.
(C2) if C_1 and C_2 are members of Ç and $C_1 \subseteq C_2$, then $C_1 = C_2$.
(C3) If C_1 and C_2 are distinct members of Ç and $e \in C_1 \cap C_2$, then there is a member C_3 of Ç such that $C_3 \subseteq (C_1 \cup C_2) - e$.

As expected, a *matroid* M is an ordered pair (E, Ç) consisting of a finite set E and a collection Ç of subsets of E satisfying conditions **(C1), (C2)** and **(C3)** above.

Alternately, we may notice certain properties of the collection I of independent sets in each example, namely

(I1) $\emptyset \in$ I
(I2) If $I \in$ I and $I' \subseteq I$ then $I' \in$ I.
(I3) If I_1 and I_2 are elements of I and $|I_1| < |I_2|$, then there is an element $e \in I_2 - I_1$ such that $I_1 \cup e \in$ I.

One of the remarkable characteristics of matroids is the several equivalent axiom systems that define them. It can be shown, see for example (6), that a set E together with a collection of subsets I which satisfy **I1 - I3** is a matroid. The rank of a matroid is defined to be the cardinality of a maximal independent set. (The reader should verify that rank is well defined.)

2. Matroid Representations

In order to gain a little more intuition, let us now look at some examples of matroids. We have seen by the example above that matrices can give rise to matroids. More formally, if E is the set of columns of an $m \times n$ matrix A over a field F, and I the set of subsets of columns that are linearly independent in the vector space V(m,F), then I is the set of independent sets of a matroid on E. The matroid obtained from a matrix in this manner is denoted M[A] and is called a vector matroid. A matroid that is isomorphic to a vector matroid over a field F is said to representable over F. There is much research activity devoted to matroid representability.

An important class of matroids arises from graphs. It can be shown that if E is the set of edges of a graph G and Ç the set of edge sets of circuits of G, then Ç is the set of minimal dependent sets of a matroid on E. For this reason, minimal dependent sets in matroids are often referred to as circuits. The matroid derived from the graph G is called the *polygon matroid* (or *cycle matroid*) of G and is denoted by M(G). A matroid that is isomorphic to the polygon matroid of a graph is called *graphic.*

Example. Let G be the complete graph K_m, the graph on m vertices with exactly one edge between each pair of distinct vertices (K_4 is pictured). $M(K_m)$ is a matroid with m(m-1)/2 elements and has rank m-1.

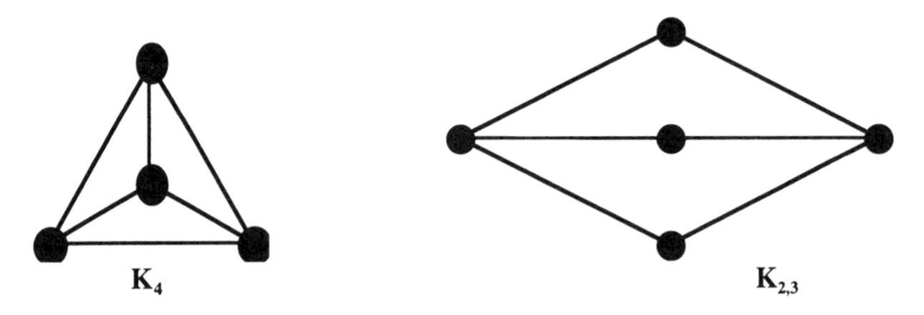

Example. A bipartite graph is one whose vertex set can be partitioned into two subsets U and V so that each edge has one end in U and one end in V. A complete bipartite graph is a bipartite graph with bipartition (U,V) in which there is exactly one edge between each vertex of U and each vertex of V. The complete bipartite graph is denoted by $K_{m,n}$, where m = |U| and n = |V|. $K_{2,3}$ is pictured above. $M(K_{m,n})$ is a matroid with mn elements and has rank m+n-1.

As it turns out, all matroids with three or fewer elements are graphic. We give a graphic representation of the eight matroids on exactly three elements in a table below, where we also suggest a matrix representation over the Reals. It is true, see for example (6), that every graphic matroid is representable over every field.

The eight non-isomorphic matroids on three elements

a graphic representation	**a matrix representation**
	$[000]$
	$[100]$
	$[110]$
	$\begin{bmatrix}100\\010\end{bmatrix}$
	$[111]$
	$\begin{bmatrix}101\\010\end{bmatrix}$

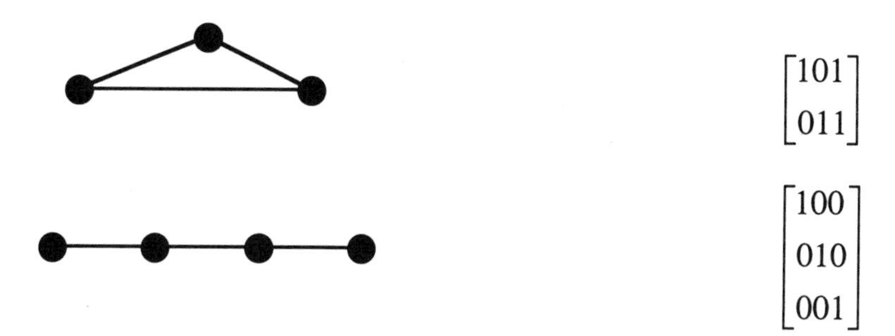

Although the matroids we are interested in are graphic, certainly not all matroids are; hence we now consider another well known class of matroids.

Let m and n be non-negative integers such that $m \leq n$. Let E be an n-element set and let the independent subsets of the matroid $U_{m,n}$ be any set with $\leq m$ elements. Note that every element of $U_{0,n}$ is a loop. $U_{1,n}$ consists of a single parallel class of elements. For $m \geq 2$, $U_{m,n}$ is simple (no loops or parallel elements). Un,n has no dependent sets, and is called a free matroid. $U_{2,4}$, commonly referred to as the four point line, is of interest in that it is not graphic and cannot be represented over GF(2).

For many other interesting examples of matroids, see the appendix in (6)

3. Independent Sets and the Unimodality Conjectures.

Let M be a matroid of rank r on a (finite) set E. Let $i_k = i_k(M)$ denote the number of independent sets of size k, $0 \leq k \leq r$. For example, if we consider $M(K_{2,3})$, we see that

$$i_0 = 1, i_1 = \binom{6}{1} = 6, i_2 = \binom{6}{2} = 15, i_3 = \binom{6}{3} = 20, i_4 = \binom{6}{4} - 3 = 12.$$

An *independent i-partition* of M is an ordered partition (A,B) of E such that A and B are independent in M and the cardinality of A is i. In a graphic representation of a graphic matroid, we are coloring the edges with two colors so that each color is a forest. We denote by $\Pi_i(M)$ the set of independent i-partitions of M and by $\pi_i = \pi_i(M)$ its cardinality. The reader can check that for $M(K_{2,3})$, we have $\pi_2 = 12, \pi_3 = 20, \pi_4 = \pi_2 = 12$.

A string of conjectures have been made about the sequences (i_k) and (π_k).

Conjecture 1. $i_k \geq \min\{ i_{k-1}, i_{k+1} \}$, $2 \leq k \leq r-1$. Welsh's unimodal conjecture, 1969

Conjecture 2. $i_k^2 \geq i_{k-1}\, i_{k+1}$, $1 \leq k \leq r-1$. Mason's log-concavity conjecture, 1972

Conjecture 3. $i_k^2 \geq ((k+1)/k)\, i_{k-1}\, i_{k+1}$, $1 \leq k \leq r-1$. Mason, 1972

Conjecture 4. $i_k^2 \geq ((k+1)/k)((m-k+1)/m-k))\, i_{k-1}\, i_{k+1}$, for $2 \leq k \leq r-1$, where $m = |E|$. Mason, '72

Conjecture 5. $\pi_{k-1} \leq \pi_k$, for $|E| = 2k$. Dowling, 1980.

Conjecture 5. $\pi_{k-1} \leq \pi_k$, for $|E| = 2k$. Dowling, 1980.

Recall that a finite sequence $a_1, a_2, \ldots, a_n$ of real numbers is said to be unimodal if $a_j \geq \min\{a_i, a_k\}$ for all i,j, and k such that $1 \leq i \leq j \leq k \leq n$. It is believed that the independent set numbers are unimodal, behaving somewhat like binomial coefficients. Recall now that a sequence is log concave if $a_k^2 \geq a_{k-1}a_{k+1}$ for all k in $\{2,3, \ldots, n-1\}$. It is an easy exercise to show that log concavity implies unimodality; and that conjectures 1 through 4 are successively stronger. Let us turn then to establishing the connection between Conjecture 5 and log concavity. Let M be a matroid of rank r. In conjecture 2, we are interested in establishing k for which $i_k^2 \geq i_{k-1}\, i_{k+1}$. In (2), Dowling defines a polynomial $f_k = f_k(M)$ over the Reals defined by

$$f_k = \sum_A \Big(\prod_{x_i \in A} x_i\Big),$$

where the sum is extended over all independent sets of size k in M. He then defines a homomorphism σ: R $\rightarrow$ Z, under which the image of a polynomial is the sum of its coefficients, and establishes that

$$i_k^2 = \sigma^2(f_k) = \sigma(f_k^2) \geq \sigma(f_{k-1}f_{k+1}) = \sigma(f_{k-1})\,\sigma(f_{k+1}) = i_{k-1}\, i_{k+1}.$$

Interpreting the coefficients in the homogeneous polynomials f_k^2 and $f_{k-1}f_{k+1}$ leads one to the independent partitions $\Pi_n(N)$ for certain minors of the matroid M. Dowling then proves the following result connecting independent partitions to log concavity:

Theorem 1. Let M be a class of (finite) matroids closed under minors and such that for every $n \leq k$ and every N in M with 2n elements we have $\pi_{n-1}(N) \leq \pi_n(N)$, then $i_k^2 \geq i_{k-1}\, i_{k+1}$ for every M in M.

A discussion of matroid minors is beyond the scope of this paper; for a detailed discussion of matroid minors, the reader is referred to (6). Of course, the class of all matroids is closed under minors. Hence, if one can prove Theorem 1 for the class of all matroids, then one would have established that the independent set numbers of matroids are log concave.

4. Conclusion

Let us now turn to a survey of known results related to the conjectures listed above [6]. In 1972, Mason (5) proved that $i_k \leq i_{r-k}$ for $k \leq r/2$. In 1975, Seymour (7) proved that Conjecture 3 is true for matroids M in which there is a k such that any circuit C of M has $|C| = 1$, $|C| = 2$, or $|C| \geq k$. In 1989, Hamidoune and Salaun (3) proved Conjecture 4 for k = 3 and conjectured another invariant of it. In 1980, Dowling (2) proved Conjecture 5 for $k \leq 7$, for the class of all matroids. The question of the log concavity of the independent set numbers for arbitrary matroids with at least sixteen elements is still open.

Using Theorem 1, some progress has been made in establishing these conjectures for certain well behaved (minor closed) classes of matroids. In 1985, Mahoney (4) proved Conjecture 5 for polygon matroids of outerplanar graphs. And Anthony Nance, in his doctoral thesis at Ohio State to appear in 1997, has proved Conjecture 1 for the

5. References

1. J. A. Bondy and U.S.R.Murty, *Graph Theory with Applications*, North Holland, New York, 1976.
2. T. A. Dowling, *On the Independent Set Numbers of a Finite Matroid,* Annals of Discrete Mathematics **8** (1980) 21-28.
3. Y. O. Hamidoune and I. Salaun, *On the Independence Numbers of a Matroid,* J. Combinatorial Theory, Series B, **47,** 146-152.
4. Carolyn R. Mahoney, *On the Independent Set Numbers of a Class of Matroids,* Journal of Combinatorial Theory, Series B, **39**, (1985) 77-85.
5. J. H. Mason, *Matroids: Unimodal Conjectures and Motzkin's Theorem,* in Combinatorial Math. Inst. Oxford, 1969, pp. 164-203. (D.J.A. Welsh and D. R. Woodal, Eds.) Academic Press, Lond/New York, 1971.
6. James G. Oxley, *Matroid Theory,* Oxford Science Publications, New York, 1992
7. Paul D. Seymour, *Matroids, hypergraphs and the max-flow min-cut theirem.* D. Phil. thesis, University of Oxford.

DIMACS Series in Discrete Mathematics
and Theoretical Computer Science
Volume **34**, 1997

On Achieving Channels in a Bipolar Game

Curtis Clark

ABSTRACT. Let m, n and r be positive integers. The (n,r,m)-bipolar game has two players A and B. The vertices of the game board consist of two labeled vertices u and v separated by r mutually disjoint junctions, J_1, J_2, J_3, ..., J_r, each of which contains m vertices. The edges of the game board consist of all edges from u to the vertices in J_1, all the edges from v to the vertices in J_r, and all edges between consecutive junctions. A channel is a path from u to v which contains exactly one vertex in each junction. Two channels are distinct if they have only u and v in common. With Player A going first, the players alternately color different edges green and red respectively. The goal of Player A is to make n distinct channels in his color. The goal of Player B is to prevent A from making n such channels, or, if prevention is not possible, he makes his moves so that Player A takes the most number of moves before winning. The minimum m such that Player A wins the (n,r,m)-bipolar game is denoted bc(n,r). We show that bc(n,2) = 2n+1 and that bc(n,r) = 2n or 2n+1 for $r \geq 3$.

1. Preliminaries

Let F and K be graphs with no isolated vertices. Notation and terminology follow [3], unless otherwise stated. The *F-achievement game on K* has two players A and B who alternately color the edges of K [1]. Player A goes first. The first player who makes a copy of F in his color is the winner; otherwise, the play is a draw. From game theory, we know that this is a first player win game. If Player A wins the F-achievement game on K, then F is *achievable* on K. The graph F is *economical* on K if A can win using the same number of moves as there are edges in F. The graph F is *ultimately economical* (u.e.) if there exists K on which F is economical. The F-achievement game on a complete graph is defined in [2].

In a play of the F-achievement game on K, Player A has a (single) *threat* to make F after his move if there is exactly one uncolored edge e such that F can be made on his next move. Of course, since we are assuming rational play, Player B will always

---

1991 *Mathematics Subject Classification.* Primary 05C99; Secondary 05C15

block such a threat. Player A has a *double threat* (d.t.) to make F after his move if there are at least two edges e_1 and e_2 such that Player A can make F on his next move by coloring either e_1 or e_2 . If Player A has a d.t. then Player B cannot stop him from winning the play of the game.

In a previous work, we proved the following theorem [1].

Theorem 1.1: *Every graph F with no isolated vertices has a supergraph which is ultimately economical.*

For positive integers, n, m and r, the *(n,r,m)-bipolar game* is described as follow. There are two players A and B. The vertices of the game board consist of two labeled vertices u and v, called *poles,* separated by r mutually disjoint junctions, J_1, J_2, J_3, ..., J_r, each of which contains m vertices. The edges of the game board consist of all edges from u to the vertices in J_1, the *u-junction*, all the edges from v to the vertices in J_r ,the *v-junction*, and all edges between consecutive junctions. The junctions J_1 and J_r are the *polar junctions*. A *channel* is a path from u to v which contains exactly one vertex in each junction. *Distinct channels* have only u and v in common. With Player A going first, the players alternately color different edges green and red respectively. The goal of Player A is to make n distinct channels in his color. The goal of Player B is to prevent A from making n such channels, or, if prevention is not possible, he makes his moves so that Player A takes the most number of moves before winning.

The (n,1,m)-bipolar game is second player win for all m. Player B blocks each single threat by coloring the edge to the other pole. The methods used in proving Theorem 1.1 can be used to show that for r > 1, there exists m such that the (n,r,m)-bipolar game is first player win. The minimum m such that Player A wins the (n,r,m)-bipolar game is the *(n,r,m)-bipolar number* and is denoted bc(n,r). In the (n,r,m)-bipolar game, a player *captures* a vertex in a polar junction if he colors an edge from a pole to that vertex.

The complete bipartite $K_{s,t}$ graph will be denoted will be denoted K(s,t). For a graph G, the notation G - te is used to denote any one of the graphs obtained from G by deleting t edges. If K = G - te, the *deleted degree* of a vertex is the number of deleted edges incident with the vertex. The following result will be needed.

Lemma 1.2: *If F is achievable on G - te, $t \geq 1$, then F is achievable on G - pe, $0 \leq p < t$.*

Proof: Since Player A has a winning strategy for achieving F on G - te, he has a winning strategy on G - pe by playing as if t - p more edges are deleted. //

2. Main Results

The following results which show that n independent edges, nK_2, are economical on certain complete bipartite graphs with deleted edges will be used.

Lemma 2.1: *The graph K_2 is economical on K(1,2) - e and for $n \geq 2$, the graph nK_2 is economical on K(n,n+1) - te, $0 \leq t \leq n-1$.*

Since K(1,2) - e has exactly one edge, Player A makes K_2 by coloring this edge on his first move. Thus, K_2 is economical on K(1,2) - e. By Lemma 1.2, to prove that nK_2 is economical on K(n,n+1) - te, $0 \leq t \leq n-1$, $n \geq 2$, it is enough to prove that nK_2 is economical on K(n,n+1) - (n - 1)e, $n \geq 2$. We use mathematical induction.

Consider the $2K_2$ -achievement game on K(2,3) - e. Label the two parts of K(2,3) as J_1 and J_2 respectively. Label the vertices of J_1 as u_1 and u_2 and the vertices J_2 as v_1, v_2 and v_3. Suppose u_1v_1 is the deleted edge (any other deleted edge is equivalent to this one). Player A colors u_1v_2 and has a double threat to make $2K_2$ with edges u_2v_1 and u_2v_3. Thus, that $2K_2$ is economical on K(2,3) - e. By Lemma 1.2, $2K_2$ is economical on K(2,3).

Assume for some $n \geq 2$ that nK_2 is economical on K(n,n+1) - te, $0 \leq t \leq n-1$. Consider the $(n+1)K_2$ achievement game on K(n+1,n+2) - ne. Either there is a vertex on the game board with deleted degree greater than or equal to two or every vertex on the game board has deleted degree zero or one. Again, we label the two parts of K(n,n+1) as J_1 and J_2 respectively.

Case 1: There is a vertex with deleted degree ≥ 2. Let w be a vertex with maximum deleted degree. Since the deleted degree is at most n there is at least one vertex u in J_1 and at least two vertices v_1 and v_2 in J_2 which have deleted degree zero. If $w \in J_1$, then Player A colors wv_1. If $w \in J_2$, then Player A colors u_1w. In either case, at least two deleted edges will not join the uncovered vertices in the two parts. Now, by treating the next move of Player B as a deleted edge, the remaining strategy of Player A becomes that of making nK_2 on K(n,n+1) - te, for some t, $0 \leq t \leq n-1$. This can be done by assumption.

Case 2: Every vertex has deleted degree zero or 1. Let u_1v_1 and u_2v_2 be deleted edges. Player A colors u_1v_2. Now, these two deleted edges do not join uncovered vertices in the two parts. As in the above case, strategy of Player A becomes that of the hypothesis of the induction argument.

In both cases, we have shown that $(n+1)K_2$ is economical on K(n+1,n+2) - ne, $n \geq 2$. It follows that $(n+1)K_2$ is economical on K(n+1,n+2) - te, $0 \leq t \leq n$. Thus, by mathematical induction, for $n \geq 2$, nK_2 is economical on K(n,n+1) - te, $0 \leq t \leq n-1$.//

Lemma 2.2: *If $n \geq 1$, then the graph nK_2 is economical on K(n,p) - te, $p \geq n+2$ and $0 \leq t \leq p-1$.*

Proof: By Lemma 1.2, it is enough to show that nK_2 is economical on K(n,p) - (p-1)e, $p \geq n+2$. We give an induction argument to complete the proof.

If n = 1 and $p \geq 3$, the K_2 is economical on K(1,p) - (p -1)e because exactly one edge remains after deleting the p -1 edges. Player A colors this edge. Thus, K_2 is economical on K(1,p) - (p -1)e and, hence, K_2 is economical on K(1,p) - te, $0 \leq t \leq p-1$.

Assume for some $n \geq 1$, the graph nK_2 is economical on K(n,p) - te, $p \geq n+2$ and $0 \leq t \leq p-1$. Consider the nK_2 achievement game on K(n+1,p) - (p-1)e, $p \geq n+3$. Again label the parts of the game board as J_1 and J_2 respectively. Since $p-1 \geq n+2$, there must be at least one vertex u in J_1 with deleted degree ≥ 2. Also, since there are only p-1 deleted edges, there must be at least one vertex v in J_2 with deleted degree equal to zero. Player A colors the edge uv. Now, at least two deleted edges will not join the uncovered vertices in the two parts. By treating the next move of Player B as a deleted edge, the remaining strategy of Player A becomes that of making nK_2 on $K(n,p) - te$, $p \geq n+2$ and $0 \leq t \leq p-1$, for some t. This can be done by assumption. Thus, $(n+1)K_2$ is economical on $K(n+1,p) - (p-1)e$, $p \geq n+3$. Hence, $(n+1)K_2$ is economical on $K(n+1,p) - te$, $p \geq n+3$ and $0 \leq t \leq p-1$. By mathematical induction, we have proved that for $n \geq 1$, the graph nK_2 is economical on $K(n,p) - te$, $p \geq n+2$ and $0 \leq t \leq p-1$.//

Lemma 2. 3: *For $n \geq 1$, the graph nK_2 is economical on $K(n+1,n+1) - te$, $0 \leq t \leq n+2$.*

Proof: By Lemma 1.2, it is enough to show that nK_2 is economical on $K(n+1,n+1) - (n+2)e$. We use mathematical induction. If $n = 1$, K_2 is economical on $K(2,2) - 3e$ because one edge remains after three edges are deleted from $K(2,2)$. Thus, K_2 is economical on $K(2,2) - te$, $0 \leq t \leq 3$.

Assume, for $n \geq 1$, that the graph nK_2 is economical on $K(n+1,n+1) - te$, $0 \leq t \leq n+2$. Consider the $(n+1)K_2$ achievement game on $K(n+2,n+2) - (n+3)e$. Let J_1 and J_2 be the two parts of the bipartite game board. Since n+3 edges are deleted there must be at least one vertex u in J_1 or J_2 with deleted degree ≥ 2. If $u \in J_1$ then there is at least one vertex v in J_2 which does not have a deleted edge in common with u. Player A colors the edge uv. Now, at least two deleted edges will not join the uncovered vertices in the two parts. By treating the next move of Player B as a deleted edge, the remaining strategy of Player A becomes that of making nK_2 on $K(n+1,n+1) - te$, $0 \leq t \leq n+2$, for some t. This can be done by assumption. Thus, $(n+1)K_2$ is economical on $K(n+2,n+2) - (n+3)e$, and, hence, $(n+1)K_2$ is economical on $K(n+2,n+2) - te$, $0 \leq t \leq n+3$. We conclude by mathematical induction that for $n \geq 1$, the graph nK_2 is economical on
$K(n+1,n+1) - te$, $0 \leq t \leq n+2$.//

Theorem 2.4: *The (n,2)-bipolar number is 2n +1.*

Proof: We do this proof in two parts.

Part I: The (n,2)-bipolar number is > 2n. Consider the (n,2,2n)-game. Player B captures n vertices in each polar junction with his first 2n moves. If Player A is to make n channels, then he is forced to capture the remaining n vertices in each polar junction. Now, Player B can color edges arbitrarily until A makes n - 1 channels. Then Player B blocks Player A from making n channels by coloring the single threat created by Player A to make the last channel. Thus, $bc(n,2) > 2n$.
Part II: The (n,2)-bipolar number is $\leq 2n + 1$. A winning strategy for Player A on (n,2,2n+1)-bipolar game is described. Player A begins by capturing as many vertices in the u-junction (J_1). He will capture at least n+1 vertices and at most 2n + 1 vertices. Then he proceeds to capture n vertices in the v-junction (J_2). If Player A captures exactly n+1 vertices in J_1 and n vertices in J_2, and Player B has make no edges between the captured vertices of Player A, then the strategy of Player A is to

make nK_2 on K(n,n+1) or K(n,n+1) - e depending on the next move of Player B. In either case, by Lemma 1.1, Player A wins.

If Player A captures exactly n+1 vertices in J_1 and Player B does color edges between these vertices, then Player B can color at most n such edges. Player A now captures another vertex in J_2. Now the strategy of Player A is to make nK_2 on K(n+1,n+1) - ne or K(n+1,n+1) - (n+1)e depending on the next move of Player B. In either case, by Lemma 2.3, Player A wins.

If Player A captures exactly n+2, n+3, ..., or 2n+1 vertices in J_2 and n vertices in J_1, then either Lemma 2.1 applies to the captured vertices or a subgraph of the captured vertices or Lemma 2.2 is applicable by Player A capturing another vertex in J_1. Thus, $bc(n,2) \geq 2n+1$.

We conclude from Parts I and II that bc(n,2) = 2n + 1.//

Theorem 2.5: *If $r \geq 3$, then $2n \leq bc(n,r) \leq 2n + 1$.*

Sketch of Proof: To show that bc(n,r) > 2n - 1, consider the (n,r,2n - 1)-bipolar game, Player B captures at least n vertices in one of the polar junctions and Player A cannot get n channels through because he can capture only n -1 vertices in this junction.

To make n channels in the (n,r,2n+1)-bipolar game, Player A begins by capturing as many vertices in the polar junctions as Player B will allow. He will capture at least n vertices in one junction and n+1 vertices in the other junction. Once this is done, similar results to Lemmas 2.1, 2.2, and 2.3, insures that there are too many options for Player A to get the n channels through for Player B to block him. Thus, bc(n,r) is 2n or 2n +1. //

We conjecture that the (n,r) bipolar number is 2n + 1 for $r \geq 3$.

3. Final Remarks

There are some interesting variations on this game. Instead of ordering the junctions, consider the same game where the intermediate junctions are unordered. Secondly, games where the first player has two moves to one move for the second player and vice versa, and other variations of this kind, should lead to some related results. Thirdly, games with more than two poles should be investigated. Applications of the theory is an open question. We anticipate its usefulness in telecommunications, cryptography and other related areas.

4. References

1. Curtis Clark, Frank Harary, and Thomas Storer, Ultimately Economical Graphs, *Congressus Numerantium*, 64 (1988), 81 - 88.
2. Frank Harary, Achievement and avoidance games on graphs. *Annals of Discrete Math.* 13 (1982) 111-120.
3. Frank Harary, *Graph Theory*, Adison-Wesley, Reading (1969).

Department of Mathematics, Morehouse College, Atlanta, GA 30314
Current Address: Dept. of Mathematics, Morehouse College, Atlanta, GA 30314
E-mail address: cuclark@morehouse.edu

DIMACS Series in Discrete Mathematics
and Theoretical Computer Science
Volume **34**, 1997

DISCRETE APPROXIMATION OF INVARIANT MEASURES FOR MULTIDIMENSIONAL MAPS

WALTER M. MILLER

ABSTRACT. In 1960 S. M. Ulam proposed a numerically feasible method for the discrete approximation of invariant measures for dynamical systems. In 1976 theoretical justification was given by T. Y. Li who proved the method's convergence for a class of expanding interval maps. Heretofore extensions only partial, or for variants of the method, have appeared in the literature. In this note we provide the mathematical and historical background leading up to a recently announced multidimensional generalization of Li's results.

0. INTRODUCTION

In 1960 S. M. Ulam ([SMU]) proposed a natural, intuitively appealing and numerically feasible method for the discrete approximation of invariant measures for dynamical systems. While in practice the method converges for a wide class of systems, theoretical justification was not forthcoming until T. Y. Li, in 1976, proved the method's convergence for a class of piecewise expanding interval maps; cf. [TYL].

It turns out that the tools needed to generalize Li's

1991 *Mathematics Subject Classification.* Primary 58F11; Secondary 28D05.

results to the multidimensional case were extant since the eighties, but heretofore extensions only partial, or for variants of the method, have appeared in the literature; cf. [B/L], [D/Z], [GF]. The goal of this note is to provide some mathematical, scientific, and historical context for the multidimensional extensions of Li's results announced in [D/Z,2] and [WMM3].

The organization of the paper is as follows. In section 1 we begin with a brief discussion of the mathematical and physical significance of invariant measures. Then in section 2 we describe Ulam's approximation method itself. In section 3 we outline the strategy behind T. Y. Li's proof of convergence of the method, and finally after a brief introduction of the requisite tools in section 4, we indicate the extension to the multidimensional case, sketching the bare bones of the proof of the main result announced in [WMM3].

1. BACKGROUND

Historically DYNAMICAL SYSTEMS THEORY arose out of the mathematical analysis of solutions of physical systems modelled by differential equations. The solutions induce FLOWS (continuous time dynamical systems) on the state, phase or event space X of the physical system, and these flows induce maps T on X, which by iteration may be regarded as discrete time dynamical systems on X. This view of T as a dynamical system is often denoted by writing $x_{n+1} = T(x_n)$, and as such, the TRAJECTORY or ORBIT $\mathcal{O}(x)$ of any $x \in X$ is defined as the sequence of iterates,

$$\mathcal{O}(x) := \{T^n x \mid \ T^n x := T(T^{n-1}x),\ T^0(x) := x,\ \forall n \geq 1\}.$$

ERGODIC THEORY involves the study of statistical properties of motions (dynamical systems) in a measure (probability) space

ERGODIC THEORY involves the study of statistical properties of motions (dynamical systems) in a measure (probability) space X. It originated at the turn of the century out of the mathematical analysis of systems in Mechanics modelled by divergence free vector fields and out of the mathematical analysis of problems in Statistical Mechanics; see eg. [DR]. It turns out that divergence free systems induce volume preserving flows and maps; ie. Lebesgue measure m is an invariant measure for the resulting flow. More precisely

Definition 1.1: A measure μ on X is an INVARIANT MEASURE for a map T if for any (measurable) subset A of X, $\mu(T^{-1}A) = \mu(A)$, where $T^{-1}A \equiv \{x: T(x) \in A\}$.

Now a measure μ on X assigns numbers to subsets of X in a manner consistent with our notion of area, volume or probability; and given a dynamical system induced by a map T on X, we seek a μ that 'measures' the statistical distribution of orbits of T. This is provided by an asymptotic measure for T which roughly speaking assigns to a subset the probability of it being visited by a typical orbit of T. Recall that an assertion (regarding x) holds *almost everywhere* w.r.t μ (μ-a.e.) if it fails on at most a subset of μ-measure zero.

Definition 1.2: A measure μ is said to be an ASYMPTOTIC MEASURE for a map T on X if for μ-a.e. x in X, and any integrable function F on X,

$$\lim_{n\to\infty} \frac{F(x) + F(Tx) + \cdots + F(T^{n-1}x)}{n} = \int_X F \, d\mu \qquad (1)$$

ie. the (time) average of any function (eg. measurement) F over a typical trajectory is equal to the space average (over X) of F.

It is a fundamental result in ergodic theory that ERGODIC T-INVARIANT MEASURES ARE ASYMPTOTIC MEASURES for T. The ergodicity of T (w.r.t. μ) is an indecomposability condition on T guaranteeing that it cannot be expressed as a 'union' of two independent dynamical sytems $T_i: X_i \to X_i$ on mutually disjoint subsets X_i of X, each of positive μ-measure, $i = 1,2$. We have

BIRKHOFF ERGODIC THEOREM (1931): Let T denote a dynamical system on a measure space (X,μ). If μ is T-invariant and if T is ergodic w.r.t. μ, then (1) holds, ie. μ is an asymptotic measure for T.

Usually (experimentally) asymptotic measures, especially those for which (1) also holds for Lebesgue a.e. x, are approximated by COMPUTING TRAJECTORIES. The supports of these measures are called ATTRACTORS as they indicate where these measures are positive and hence where 'most' orbits, in a statistical sense, are concentrated; see eg. [DR].

2. A BRIEF DESCRIPTION OF ULAM'S METHOD

If one views a measure on X, for example, as a (mass) distribution or density on X one sees that the action of a dynamical system T, since it redistributes points, necessarily induces a corresponding action P_T on measures. P_T is called the PERRON-FROBENIUS (PF) OPERATOR for T. We follow with a precise definition of P_T in terms of its effect on densities - nonnegative functions with unit integrals. Equation (2) well defines P_T whenever T is NONSINGULAR, ie. T maps no set of positive measure onto a set of measure zero; cf. [L/M].

Definition 2.1: The *Perron–Frobenius operator* P_T corresponding to a nonsingular transformation T on (X,m) is the unique linear operator on $L_1(X,m)$ such that for any (measurable) subset A of X and any (integrable) function F on X,

$$\int_A P_T F \, dm = \int_{T^{-1}A} F \, dm \tag{2}$$

Now (2) implies that a fixed density of P_T corresponds to an invariant measure μ, where $\mu(A) := \int_A F \, dm$, $A \subseteq X$. So in order to find T-invariant measures we see that it suffices to find fixed densities of P_T. This leads to alternative approaches to the usual approximation of invariant measures by computing trajectories.

In 1960 S. Ulam proposed the following scheme for appproximating invariant densities for a transformation T. For a given equipartition Π_n of the space X into $\{X_1, X_2, \ldots, X_n\}$ define a row stochastic matrix $[P_n]$, whose (i,j)-th entry gives the fraction of X_i that is mapped by T to X_j, ie.

$$[P_n]_{ij} := \frac{m(X_i \cap T^{-1}X_j)}{m(X_i)} \tag{3}$$

It turns out that the matrix $[P_n]$ corresponds to a finite approximation P_n to the PF operator, $P = P_T$. P_n is defined as follows: For a density F let $Q_n F$ denote the piecewise constant (w.r.t. Π_n) density whose value at x is the average of F on the pixel of the partition containing x,

$$Q_n F = \sum_{j=1}^{n} c_j 1_{X_j}, \quad c_j = \frac{1}{m(X_j)} \int_{X_j} F \, dm \tag{4}$$

where 1_A denotes the indicator function on a subset A. Then

$$P_n := Q_n P_T. \tag{5}$$

Now by the theory of nonnegative matrices (cf. eg. [B/P]), $[P_n]$ has at least one normalized nonnegative (left) fixed point $F_n = (a_1,\dots,a_n)$, and this corresponds to the step function

$$\sum_{k=1}^{n} a_k \cdot 1_{X_k},$$

which by 'normalized nonnegative' is seen to be a density. As such the relationship between $[P_n]$ and P_n may be described explicitly as follows,

$$(a_1,\dots,a_n)\cdot[P_n] = (b_1,\dots,b_n)$$

iff

$$P_n\Big(\sum_{k=1}^{n} a_k \cdot 1_{X_k}\Big) = \sum_{k=1}^{n} b_k \cdot 1_{X_k}.$$

Now for the sake of simplicity, let $I = [0,1] \subset \mathbb{R}$, m be Lebesgue measure on I and assume that a nonsingular transformation $T:(I,m) \to (I,m)$ is ergodic w.r.t. m. Ergodicity assures us that the here assumed P_T-invariant density F^* is unique; cf. [L/M]. Let Π_n denote a sequence of equipartitions of $[0,1]$ satisfying mesh$(\Pi_n) \to 0$, and recall that for any sequence of integrable functions $\{F_n\}$, F_n converges in L_1 (ie. converges in the mean) if there exists an $F^* \in L_1$ such that $\|F_n - F^*\|_1 \to 0$, as $n \to \infty$; where

$$\|F\|_1 := \int_X |F| \, dm$$

denotes the norm on L_1. We have

CONJECTURE (S.M. ULAM, 1960): ANY SEQUENCE OF NORMALIZED NONNEGATIVE FIXED EIGENVECTORS, F_n, OF THE MATRICES $[P_n]$, CORRESPONDING TO DENSITIES ON FINER AND FINER PARTITIONS OF I, CONVERGE IN $L_1(I,m)$ TO THE DENSITY F^* OF A T-INVARIANT MEASURE.

REMARK: One obvious numerical advantage of the above program comes from the reduction of the problem of approximating invariant measures to that of the computation of fixed eigenvectors of stochastic matrices. See eg. [CSH] for explicit examples.

3. A SOLUTION TO ULAM'S CONJECTURE

A solution to Ulam's conjecture was finally provided by T. Y. Li in [TYL] for expanding piecewise C^2 and monotonic transformations T of an interval I; see also [WMM1]. This means that there exists a partition $\mathcal{P}$ of I into finitely many subintervals I_k such that the restriction T_k of T on the interior of I_k is monotonic and C^2 (twice continuously differentiable). More precisely we have the following

THEOREM 3.1 (T.Y. Li, 1976): Let $T:[0,1] \to [0,1]$ be an ergodic piecewise C^2 monotonic map. If $J := \inf|T'| > 2$ then $\{F_n\}$ converges to the unique fixed point F^* of P_T.

The main ingredients in the proof are contained in the following four lemmas. As the notion of bounded variation plays such a crucial role, we first recall its definition.

Definition 3.1: For $f:[a,b] \to \mathbb{R}$ the VARIATION of f on $[a,b]$, denoted by $\mathrm{Var}(f,[a,b])$, is given by

$$\sup \{ \sum_{j=1}^{p} |f(t_j) - f(t_{j-1})| : a = t_0 < t_1 < \cdots < t_{p-1} < t_p = b\}$$

If $\text{Var}(f)$ is finite, then f is said to be of *bounded variation.* Now roughly speaking the variation measures the *vertical* distance (as opposed to arc length) traversed by a point as it moves along the graph of f. As such one is able to show the following

BV1) If f is differentiable on $[a,b]$, then

$$\text{Var}(f) = \int_a^b |f'(x)| \, dx.$$

BV2) If f is integrable and Π_n the standard equipartition of $[a,b]$ into n subintervals (see (4)), then $\text{Var}(Q_n f)$ corresponds to the *lateral* surface area (here, length) of the closure of the region contained under the graph of $Q_n f$, and over the *interior* (a,b) of $[a,b]$.

LEMMA 3.1 (Lasota & Yorke, 1973): Let T be a piecewise C^2 monotonic map on $[0,1]$. Then there exists $C = C(T) > 0$ such that

$$\text{Var}(\text{P}_T F) \leq (2/J)\cdot\text{Var}(F) + C\cdot\|F\|_1 \tag{6}$$

where $\|f\|_1$ denotes the L_1-norm of f.

Proof: See [L/Y] or [L/M], chapter 6. ∎

LEMMA 3.2: (Helly's Theorem): Any family of densities on $[0,1]$ of uniformly bounded variation is a relatively compact subset of L_1.

Proof: See [TYL]. ∎

LEMMA 3.3: W.r.t. a sequence of finer and finer partitions, if any sequence of fixed densities F_n of P_n strongly converge in L_1 to $F*$, then $F*$ is a fixed point of P.

Proof: See [TYL]. ■

LEMMA 3.4: Q_n is variation decreasing, ie. for any integrable F of bounded variation,

$$\mathrm{Var}(Q_n F) \leq \mathrm{Var}(F) \tag{7}$$

Proof: If F is continuous, then (7) follows from (4), the mean value theorem and definition (3.1) of bounded variation. For an argument and remaining details see [TYL] and [WMM3]. ■

Proof of Theorem 3.1: Granting that F_n is a fixed density of $P_n := Q_n P_\tau$, $n \geq 1$, we have by (7) and (6),

$$\begin{aligned}\mathrm{Var}(F_n) = \mathrm{Var}(P_n F_n) = \mathrm{Var}(Q_n P F_n) &\leq \mathrm{Var}(P F_n)\\ &\leq (2/J)\cdot\mathrm{Var}(F_n) + C.\end{aligned}$$

So if $(2/J) < 1$, then

$$\mathrm{Var}(F_n) \leq \frac{C}{1 - (2/J)}.$$

Hence by Lemma 3.2 the sequence of F_n's are relatively compact in $L_1(X,m)$, ie. any sequence of the F_n's have a convergent subsequence, the limit of which, by Lemma 3.3, is necessarily a fixed point $F*$ of P. If T is ergodic w.r.t. m, then P has at most one fixed point, cf. [L/M]; so the F_n's converge strongly to $F*$ in L_1, ie. Ulam's method converges. ■

REMARK: A similar strategy is employed in [WMM1] (cf. also [WMM2]) to demonstrate (one-iterate!) convergence of Ulam's method for a class of *nonuniformly expanding* transformations.

4. EXTENDING LI'S RESULT TO THE MULTI-DIMENSIONAL CASE

The convergence of Ulam's method or variants thereof for transformations in $\mathbb{R}^d$, $d > 1$ have been previously investigated by several authors, eg. [GK], [B/L], [D/Z], [FYH] and [GF]. However it is rather surprising that multidimensional analogues of Li's results were so long in coming. One of the longstanding impediments was the lack of an appropriate multidimensional generalization of the notion of bounded variation. This became available with the publication in 1984 of [EG], see also [WPZ]; which we now briefly describe.

Let Ω be an open set in $\mathbb{R}^d$. If $F \in C^1(\Omega)$, ie. if F is a continuously differentiable real-valued function on Ω, then (8) generalizes the well known formula (BV1) in the 1-dimensional case,

$$\mathrm{Var}(F,\Omega) := \int_\Omega |DF| \; dm \tag{8}$$

where for the C^1 case

$$\int_\Omega |DF| \; dm = \int_\Omega |\mathrm{grad}(F)| \; dm = \int_\Omega \sqrt{(\partial F/\partial x_1)^2 + \cdots + (\partial F/\partial x_n)^2} \; dm$$

More generally if $F \in L_1(\Omega,m)$, one defines

$$\int_\Omega |DF| \; dm := \sup \left\{ \int_\Omega F \, \mathrm{div}(G) \; : \; G \in C_0^1(\Omega,\mathbb{R}^d), \; \max(|G|) \le 1 \right\},$$

where $\mathrm{div}(G) := \partial G_1/\partial x_1 + \cdots + \partial G_n/\partial x_n$ denotes the divergence of G, $C_0^1(\Omega;\mathbb{R}^d)$ denotes the space of C^1 functions from Ω to $\mathbb{R}^d$ with compact support, and $\max(|G|)$ denotes the maximum value of $|G(x)|$, the standard Euclidean norm of $G(x)$, on Ω.

Let's now check, in the light of (8), if d-dimensional versions of the requisite four lemmas 3.1 through 3.4, for a prospective proof of the convergence of Ulam's method, hold as above. Regarding Lemma 3.1, following [G/B], let Ω be a rectangular region in $\mathbb{R}^d$ and $\mathcal{P} = \{\Omega_1,...,\Omega_p\}$ be a partition of Ω into a finite number of subsets having piecewise C^2 boundaries. Furthermore, where C^2 segments meet, assume that the angle subtended by tangents to these segments at the point of contact is bounded away from 0.

Let $T:\Omega \to \Omega$ be piecewise C^2 on $\mathcal{P}$ and expanding in the sense that there exists $J > 1$ such that for any $i = 1,2,..,p$, $\|DT_i^{-1}\| < 1/J$, where DT_i^{-1} is the derivative matrix of T_i^{-1}, $T_i := T|_{\Omega_i}$, and $\|\cdot\|$ is the Euclidean matrix norm. Gora and Boyarsky (1989) showed that the following d-dimensional version of the Lasota-Yorke inequality (6) holds.

LEMMA 4.1: There exists constants $A > 0$ and $K > 0$, such that for any $F \in BV(\Omega)$

$$\mathrm{Var}(P_T F) \leq \frac{1 + A}{J} \cdot \mathrm{Var}(F) + K \cdot \|F\|_1 \tag{9}$$

Proof: See [G/B]. ∎

They also note that if the defining partition $\mathcal{P}$ for T is rectangular we have $A = \sqrt{d}$, cf. [G/B]. In this case (9) is seen to directly generalize the corresponding inequality (6) above.

LEMMA 4.2: The multidimensional version of Helly's Theorem holds, ie. a family of densities of uniformly bounded d-dimensional variation is precompact in $L_1(\Omega)$.

Proof: See Theorem 1.19 in [EG]. ∎

LEMMA 4.3: If a sequence of Ulam approximating eigendensities F_n converges to F^* in $L_1(\Omega, m)$, then $F^* \in \mathrm{Fix}(P_T)$.

Proof: That $Q_nF \to F$ in $L_1(\Omega,m)$, for any integrable F, holds follows directly from Theorem IV.8.18 in [D/S]. The rest of the assertion now follows exactly as in [TYL], p. 184. ■

It turns out that the MISSING INGREDIENT is an analogue of LEMMA 3.4 regarding the variation decreasing property of the Q_n's. One is able to show that this does NOT hold in the d-dimensional case and as announced in [WMM3] we have the following

THEOREM 4.4 ([WMM3], 1996): With respect to any rectangular partition Π_n of a d-dimensional cube Ω, the corresponding sequence of piecewise constant averaging operators Q_n satisfy

$$\mathrm{Var}_d(Q_nF,\Omega) \leq \sqrt{d}\cdot\mathrm{Var}_d(F,\Omega) \tag{10}$$

Proof: For details see [WMM3]. ■

Now following the proof of Theorem 3.1, we are able to show

THEOREM 4.5: Let $T:\Omega \to \Omega$ be an ergodic piecewise C^2 expanding map as defined above. If T is sufficiently expanding (eg. if $J > \sqrt{d}\cdot(1 + A)$), then the Ulam eigendensities $\{F_n\}$ converge to the unique fixed point of P_T.

We close with a sketch in the next section of the ingredients of the proof of Theorem 4.4.

5. SKETCH OF PROOF OF MAIN THEOREM 4.4 FROM [WMM3]

Let $\Omega = I_1 \times \cdots \times I_d$ denote an open rectangular region in $\mathbb{R}^d$, where $I_k = (a_k,b_k) \subset \mathbb{R}$, are nonempty bounded open intervals, $k = 1,\dots,d$. For each k, let Π_{k,n_k} denote the standard equipartition of I_k into n_k subintervals; let $n := (n_1,\dots,n_d)$, and let $\Pi_n = \Pi_{1,n_1} \times \cdots \times \Pi_{d,n_d}$ denote the product equipartition of Ω into $n_1 \cdots n_d$ d-dimensional subrectangles R. We will call any such subrectangle a pixel of Π_n, and as before, for any Lebesgue integrable f and any $x \in R$, $Q_n f(x)$ gives the average value of f (w.r.t. Lebesgue) on R.

First we note that it suffices to show theorem 4.4 for smooth F (ie. $F \in C^\infty(\Omega)$.

Lemma 5.1: If $Q: L_1(\Omega) \to L_1(\Omega)$ is a bounded linear map such that $Q(L_1(\Omega)) \subseteq BV(\Omega)$, and for some $\beta \geq 0$,

$$\mathrm{Var}(QF) \leq \beta \cdot \mathrm{Var}(F), \tag{*}$$

holds for all $F \in C^\infty(\Omega)$, then (*) holds for all $F \in BV(\Omega)$.

Then we proceed to show the smooth case by induction on the dimension d. Note that inequality (7) establishes the basis for induction. Consideration of the one-dimensional case (BV2) motivates the next key observation.

Lemma 5.2: (Geometric Representation of $\mathrm{Var}(Q_n F)$): $\mathrm{Var}(Q_n F)$ represents the lateral d-dimensional 'surface' measure of the closure of the $(d+1)$-dimensional region under the graph of $Q_n F$ in $\Omega \times \mathbb{R}$, *not* including that part over the boundary of Ω.

Now let $\mathrm{int}(A)$ denote the interior of a subset A, and let

$$\Omega(j,k) = I_1 \times \cdots \times I_{j-1} \times \mathrm{int}(I_{j,k}) \times I_{j+1} \times \cdots \times I_d,$$

where $I_{j,k}$ is the k-th subinterval of the equipartition Π_{j,n_j} of I_j. One is able to show that lemma 5.2 implies

Lemma 5.3: For $d > 1$,

$$\mathrm{Var}_d(Q_nF,\Omega) = \frac{1}{d-1} \sum_{j=1}^{d} \sum_{k=1}^{n_j} \mathrm{Var}_d(Q_nF,\Omega(j,k)).$$

The next lemma relates d-dimensional variation, in particular the summands in lemma 5.3, to their $(d-1)$-dimensional analogues, thus allowing us to apply the induction hypothesis. To this effect we view

$$\hat{\Omega}_i := I_1 \times \cdots \times \hat{I}_i \times \cdots \times I_d$$

as $(d-1)$-dimensional cross sections of Ω, infinitesimally thin versions of the $\Omega(i,j)$'s, (here the carat denotes the deletion of the indicated entry), and for $t \in I_i$ let $f_{i,t}: \hat{\Omega}_i \to \mathbb{R}$ be given by

$$f_{i,t}(y) := f(y_1,\ldots,y_{i-1},t,y_{i+1},\ldots,y_d).$$

Finally let $Q_{\hat{n}_i}$ denote the piecewise constant averaging operator on $\hat{\Omega}_i$ w.r.t. the partition of $\hat{\Omega}_i$ given by

$$\Pi_{\hat{n}_i} := \Pi_1 \times \cdots \times \hat{\Pi}_i \times \cdots \times \Pi_d,$$

where again the carat denotes the deletion of the indicated entry. Consideration of the two-dimensional case motivates the following 'Fubini type' result.

Lemma 5.4: W.r.t. the notation above, the following holds:

$$\mathrm{Var}_d(Q_nF,\Omega(j,k)) \leq \int_{I_{j,k}} \mathrm{Var}_{d-1}(Q_{\hat{n}_j}F_{j,t};\hat{\Omega}_j)\, dt.$$

Finally, the following simple calculus lemma ensures that the constants in the d-dimensional cases of (10) turn out as claimed.

Lemma 5.5: For $d > 1$, and $1 \le i \le d$, we have

$$\sum_{i=1}^{d} \sqrt{(x_1)^2 + \cdots + (x_{i-1})^2 + (x_{i+1})^2 + \cdots + (x_d)^2}$$

$$\le \frac{d\sqrt{d-1}}{\sqrt{d}} \cdot \sqrt{(x_1)^2 + \cdots + (x_i)^2 + \cdots + (x_d)^2}\,.$$

SKETCH OF PROOF OF THEOREM 4.4: By induction we first assume

$$\mathrm{Var}_{d-1}(\hat{Q_{n_j}} F_{j,t};\hat{\Omega}_j) \le \sqrt{d-1}\ \mathrm{Var}_{d-1}(F_{j,t};\hat{\Omega}_j) \qquad (*)$$

for some $d > 1$. We will indicate how (10) follows. Now an application of (*), lemma 5.3 and lemma 5.4 imply

$$\mathrm{Var}_d(Q_n F,\Omega) \le \frac{\sqrt{d-1}}{d-1} \sum_{j=1}^{d} \sum_{k=1}^{n_j} \int_{I_{j,k}} \mathrm{Var}_{d-1}(F_{j,t};\hat{\Omega}_j)\ dt.$$

Hence using the equality following (8), we have

$$\mathrm{Var}_d(Q_n F,\Omega) \le \frac{\sqrt{d-1}}{d-1} \sum_{j=1}^{d} \int_{I_j} \mathrm{Var}_{d-1}(F_{j,t};\hat{\Omega}_j)\ dt$$

$$= \frac{\sqrt{d-1}}{d-1} \sum_{j=1}^{d} \int_{I_j} \int_{\hat{\Omega}_j} \|\mathrm{grad}(F_{j,t})\|\ dy\ dt$$

$$= \frac{\sqrt{d-1}}{d-1} \sum_{j=1}^{d} \int_{\Omega} \sqrt{\left(\frac{\partial F}{\partial x_1}(x)\right)^2 + \cdots + \widehat{\left(\frac{\partial F}{\partial x_j}(x)\right)^2} + \cdots + \left(\frac{\partial F}{\partial x_d}(x)\right)^2}\ dx$$

$$\leq \frac{d\sqrt{d-1}}{\sqrt{d}} \frac{\sqrt{d-1}}{d-1} \cdot \int_{\Omega} \|\text{grad } F\| \, dx = \sqrt{d} \, \text{Var}(F,\Omega),$$

as reqired, via an application of lemma 5.5. Again the carat (^) denotes the deletion of the indicated entry. ■

REMARK: We end by noting that the proof of Lemma 5.5 suggests the suitability of

$$F(x_1,\ldots,x_d) = x_1 + \cdots + x_d$$

for verifying that inequality (10) is sharp.

6. REFERENCES

[B/P] A. Berman & R. J. Plemmons, "Nonnegative Matrices in the Mathematical Sciences", Academic Press, San Diego (1979); reissued by SIAM, Philadelphia (1994).

[B/L] A. Boyarsky & Y. S. Lou, Approximating measures invariant under higher dimensional chaotic transformations, *J. Approx. Theory* **65** (1991), 231-244.

[D/Z] J. Ding & A. Zhou, Piecewise linear Markov approximations of Frobenius-Perron operators associated with multi-dimensional transformations. *Nonlinear Analysis: Theory, Methods & Applic.* **25** (1994), 399-408.

[D/Z,2] J. Ding & A. Zhou, Finite approximations of Frobenius-Perron operators. A solution of Ulam's conjecture to multidimensional transformations, Physica D **92** (1996), 61 - 68.

[D/S] N. Dunford & J. Schwartz, "Linear Operators, Part One: General Theory", J. Wiley & Sons New York (1988).

[GF] G. Froyland, Finite approximation of Sinai-Bowen-Ruelle measures for Anosov systems in two dimensions, preprint.

[EG] E. Giusti, "Minimal Surfaces and Functions of Bounded Variation", Birkhauser, Boston (1984).

[G/B] P. Gora & A. Boyarsky, Absolutely continuous invariant measures for piecewise expanding C^2 transformations in $\mathbb{R}^n$, *Israel J. Math.* **67** (1989), 272-286.

[CSH] C. S. Hsu, "Cell-to-Cell Mapping: A Method of Global Analysis for Nonlinear Systems", Springer-Verlag (1987).

[FYH] F. Y. Hunt, Approximating the invariant measures of randomly perturbed dissipative maps, *J. Math. Analysis and Applic.* **198** (1996), 534-551.

[GK] G. Keller, Stochastic perturbations of some strange attractors, Lect. Notes in Physics, **179**, Dyn. Sys. and Chaos, 192-193 (1982).

[L/M] Lasota A. & Mackey M., *Probabilistic Properties of Deterministic Systems*, Cambridge University Press, (1985).

[L/Y] A. Lasota & J. A. Yorke, On the existence of invariant measures for piecewise monotonic transformations, *Trans. Amer. Math. Soc.* **186** (1973), 481-488.

[TYL] T. Y. Li, Finite approximation for the Frobenius-Perron operator. A soluton to Ulam's conjecture. *J. Approx. Theory* **17** (1976), 177-186.

[WMM1] W. M. Miller, Stability and approximation of invariant measures for a class of nonexpanding transformations, *J. Nonlinear Analysis: Thy. Methods and Applic.* **23** (1994), 1013-1025.

[WMM2] W. M. Miller, Existence and approximation of asymptotic measures for piecewise convex interval maps - an extended solution to Ulam's conjecture, preprint.

[WMM3] W. M. Miller, Bounded variation and finite approximation for the Frobenius-Perron operator. A multidimensional solution of Ulam's conjecture, preprint.

[DR] D. Ruelle, "Chaotic Evolution and Strange Attractors", Cambridge University Press, New York (1989).

[SMU] S. M. Ulam, "A Collection of Mathematical Problems", Interscience Tracts in Pure and Applied Math., No. 8, Interscience, New York (1960).

[WPZ] W. P. Ziemer, "Weakly Differentiable Functions", Springer Verlag, New York (1989).

MATHEMATICS DEPARTMENT, HOWARD UNIVERSITY, WASHINGTON, D.C. 20059

Current Address: Mathematics Department, University of Maryland, College Park, MD 20742

DIMACS Series in Discrete Mathematics
and Theoretical Computer Science
Volume **34**, 1997

Some numerical methods for a maximum entropy problem

Nathaniel Whitaker

ABSTRACT. We describe a numerical method developed to treat the statistical equilibrium model of coherent structures in two-dimensional turbulence. We explain the statistical model for which the method was designed to solve. A convergence proof is given along with several alternative methods for the same problem. The solution of this problem requires maximizing a nonlinear functional subject to nonlinear constraints.

1. Introduction

The equations which describe two-dimensional inviscid fluid flow are the Euler equations. As time evolves these flows become highly chaotic and turbulent on increasingly smaller and smaller spatial scales. Traditional numerical methods applied to this time dependent problem are limited by the requirement of resolving these small scales. Recently, a statistical theory for the equilibrium solution of these equations has been proposed using methods from statistical mechanics. This statistical equilibrium solution for the Euler equations is obtained by maximizing a nonlinear functional characterizing an entropy subject to the natural constraints of the flow. One of the constraints is also nonlinear making this a nontrivial optimization problem. In [**8**] and [**9**], we develop an accurate and highly efficient iterative algorithm for solving this optimization problem. If one attacked this optimization problem in the usual way numerically, an enormous system would have to be solved at each iterate and there is no guarantee that the algorithm would converge. By exploiting the structure of the optimization problem, we are able to solve a simple nonlinear problem at each iterative step. We prove that our algorithm converges globally. In [**8**] and [**9**], we demonstrate the correctness of the statistical theory along with some failings by comparing our results with numerical results from traditional methods applied to the time dependent problem. We observe such classical phenomenon as the rollup of periodic vortex layers and the merger of patches of vorticity.

In this paper, we explain the method given in [**8**] along with 2 other methods proposed for the first time here. This paper is organized as follows. In section 2, we give the equations describing the flow of a viscous incompressible fluid, the Navier-Stokes equations. In section 3, we show how the Euler equations, which describe the flow of an inviscid, incompressible fluid, are obtained. In section 4, we give the

1991 *Mathematics Subject Classification.* Primary 49M99, 65K10, 76C05; Secondary 76F10 82B80.

Partially supported by the National Science Foundation under Grant DMS-9307914.

conserved quantities of the flow. In section 5, we explain the statistical theory of Miller-Robert which predicts the most probable flow to the Euler equations and associates this flow with the equilibrium solution. In section 6, we present our numerical method to solve the problem formulated by the statistical theory. In section 7, we present an equivalent simple dual problem which we actually solve. In section 8, we discuss the convergence proof for the algorithm and in section 9, we present other algorithms given for the first time.

2. Navier-Stokes Equations

The two-dimensional flow of an incompressible fluid is described by the Navier-Stokes equations

$$\frac{\partial \vec{v}}{\partial t} + \vec{v} \cdot \nabla \vec{v} + \nabla p = \nu \triangle \vec{v}, \tag{2.1}$$

$$\frac{\partial v_1(\vec{x},t)}{\partial x_1} + \frac{\partial v_2(\vec{x},t)}{\partial x_2} = 0 \tag{2.2}$$

where $\vec{v} = (v_1(\vec{x},t), v_2(\vec{x},t))$ and $p = p(\vec{x},t)$ are the velocity and pressure fields and $\nabla = (\partial/\partial x_1, \partial/\partial x_2)$ with $\vec{x} = (x_1, x_2)$. The viscosity is given by ν and $\triangle$ is the standard Laplacian. These equations hold within a domain $\Omega \subseteq R^2$, with the boundary condition $n \cdot \vec{v} = 0$ imposed on $\partial\Omega$, where n denotes the outward unit normal field. It is widely recognized now, from both the theoretical and computational points of view, that the salient features of this fluid dynamics are more concisely described by the vorticity field $\omega(\vec{x},t)$ defined by

$$\omega(\vec{x},t) = \frac{\partial v_2}{\partial x_1} - \frac{\partial v_1}{\partial x_2} \tag{2.3}$$

Incompressiblity (2.2) implies that there exists a streamfunction $\psi(\vec{x},t)$ such that,

$$\vec{v} = \left(\frac{\partial \psi}{\partial x_2}, -\frac{\partial \psi}{\partial x_1} \right)$$

and combining that with (2.3) above, we have that

$$\omega(\vec{x},t) = -\frac{\partial^2 \psi}{\partial x_1 x_1} - \frac{\partial^2 \psi}{\partial x_2 x_2} = - \triangle \psi(\vec{x},t)$$

Equation (2.1) can be written in terms of the vorticity and velocity only and is given by

$$\frac{\partial \omega(\vec{x},t)}{\partial t} + \vec{v} \cdot \nabla \omega(\vec{x},t) = \nu \triangle \omega(\vec{x},t) \tag{2.4}$$

$$\omega(\vec{x},0) = \omega_0(\vec{x}) \tag{2.5}$$

$$\psi = 0 \text{ on } \partial\Omega \tag{2.6}$$

instead of the primitive fields $\vec{v}$ and p.

Equation (2.4) describes the transport of the vorticity. Numerically, these equations are difficult of solve especially for a small viscosity ν. As the viscosity decreases, smaller scales become more and more important and a finer grid is needed to approximate the equations numerically. For a reasonable viscosity, one can easily overwhelm the most powerful computers.

3. Euler Equations

For $\nu = 0$, equations (2.4)-(2.6) become the Euler Equations,

$$\frac{\partial \omega(\vec{x},t)}{\partial t} + \vec{v} \cdot \nabla \omega(\vec{x},t) = 0 \tag{3.1}$$

$$\omega(\vec{x},0) = \omega_0(\vec{x})$$

$$\psi = 0 \text{ on } \partial\Omega$$

with

$$\omega(\vec{x},t) = - \triangle \psi(\vec{x},t)$$

These equations are almost impossible to solve numerically because of the infinitely small scales involved, nevertheless, they have a simple interpretation. Given a vorticity particle trajectory $(t, \vec{x}(t))$, equation (3.1) can be written as

$$\frac{d\omega(\vec{x}(t),t)}{dt} = 0.$$

The means that the value of the vorticity is just transported along particle paths, i.e., $\omega(\vec{x},t)$ is just a rearrangement of $\omega(\vec{x},0)$ for $t > 0$. In our presentation, we consider only so-called vortex patches, i.e., $\omega(\vec{x},0) = 0$ or $\omega(\vec{x},0) = 1$. Consequently, as time goes on the vorticity $\omega(\vec{x},t)$ takes on the same values 0 and 1. However $\mid \nabla\omega(\vec{x},t) \mid$ grows rapidly in t, such that the patches develop smaller and smaller scale fluctuations. Any numerical method is limited by the higher and higher resolution required as time goes on.

4. Conserved Quantities

Certain quantities are invariant in the flow. The initial circulation Γ_0 at time 0 is given by,

$$\Gamma_0 = \int_\Omega \omega(\vec{x},0)d\vec{x}$$

and the initial energy E_0 at time 0 is given by,

$$E_0 = \frac{1}{2}\int_\Omega \omega(\vec{x},0)\psi(\vec{x},0)d\vec{x}.$$

As ω evolves, the circulation Γ and the energy E are conserved for all time, i.e.,

$$\Gamma = \int_\Omega \omega(\vec{x},t)d\vec{x} = \Gamma_0$$

and

$$E = \frac{1}{2}\int_\Omega \omega(\vec{x},t)\psi(\vec{x},t)d\vec{x} = E_0.$$

This can be verified by differentiating the quantities E and Γ with respect to time and using the differential equation. There are other conserved quantities depending on the geometry of Ω. In **[9]**, we find the equilibrium solution in a disk, which has the additional conserved quantity

$$M = \int_\Omega \omega(\vec{x},t)\frac{|\vec{x}|}{2}d\vec{x}.$$

In [8], we find the equilibrium solution in a x_1-periodic domain with dirichlet boundary conditions in x_2. Here,

$$I = \int_\Omega \omega(\vec{x}, t) x_2 d\vec{x}$$

is also conserved.

5. Miller-Robert Theory

The Euler equations are difficult to solve therefore we will try to use some ideas from statistical mechanics. Suppose that we start with some initial flow or initial vorticity $\omega_0(\vec{x})$ and wish to find the equilibrium vorticity distribution. Let the

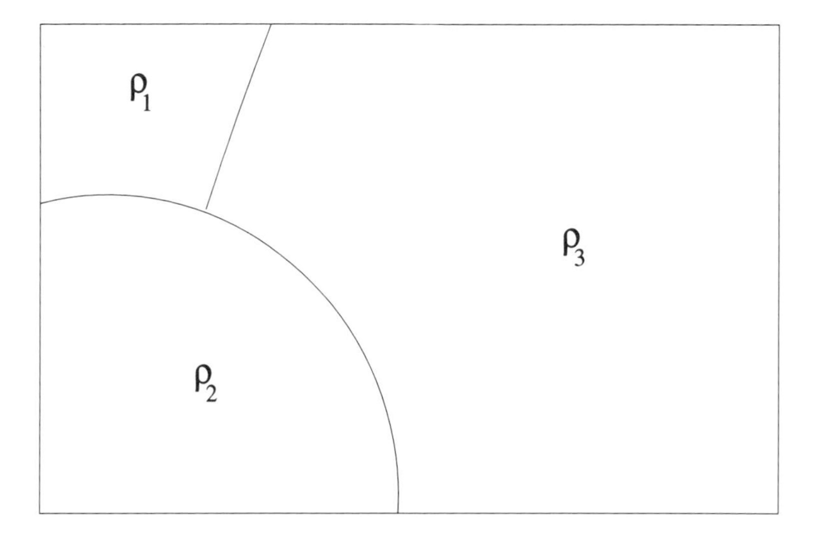

FIGURE 1. This represents all possible microstates partitioned.

region in figure 1 represent all the vorticity functions with the same circulation and energy of our given vorticity function $\omega_0(\vec{x})$. These admissible functions represent microscopic variables. We attempt to partition and associate with a subset of microscopic variables, a macroscopic variable $\rho(\vec{x})$. In figure 1, for the purpose of clarity, we assume there are only 3 possible macroscopic variables. Based on figure 1, we might guess that our equilibrium solution lies in the class described by $\rho_3(\vec{x})$. We now explain the partitioning strategy given proposed by Miller and Robert in a series of papers([2], [3], [4], [5]). We attempt to find the macroscopic variable which contains the most microscopic variables. We then hypothesize that most probably our equilibrium solution lies in this class. Robert argues that the macrostate which contains the most microstates contains overwhelmingly more than any other.

We give a heuristic derivation of the Miller-Robert theory which connects the microstates to the macrostates. We suppose that our domain Ω contains only 2 points X_1 and X_2 divided into N equal parts each where ω is 0 or 1 on each part.

Figure 2 gives a possible microstate. If we think of it's macrostate ρ as representing the local volume fraction at each point, then it is given by

$$\rho(X_1) = \frac{4}{9}, \ \ \rho(X_2) = \frac{5}{9}$$

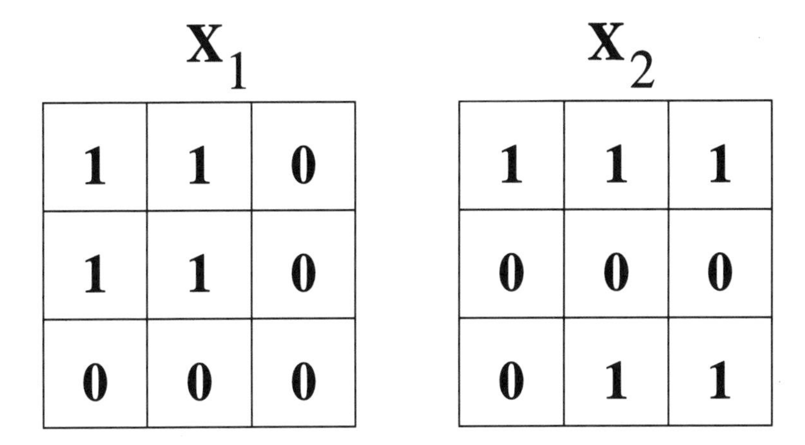

FIGURE 2. This represents our domain with 2 points.

which contains the microstate in figure 2. In fact, ρ contains

$$\frac{9!}{4!(9-4)!}\,\frac{9!}{3!(9-3)!}$$

microstates. If $\rho(X_1) = \frac{N_1}{N}$ and $\rho(X_2) = \frac{N_2}{N}$ then the number of microstates contained by this macrostate is

$$W(\rho) = \frac{N!}{N_1!(N-N_1)!}\,\frac{N!}{N_2!(N-N_2)!}.$$

Note also that $\rho(\vec{x})$ can be thought of as the probability that ω takes on the value 1 at $\vec{x}$. For our domain Ω the number of microstates associated with a macrostate is given by

$$W(\rho) = \prod_i \frac{N!}{N_i!(N-N_i)!}$$

where this product is taken over all the points in Ω. After some simplification, one finds that the macrostate ρ with the most microstates is the one which maximizes

$$S(\rho) = -\int_\Omega \rho\log(\rho) + (1-\rho)\log(1-\rho)d\vec{x}.$$

We can relate our macroscopic variable back to our microscopic variables. The expected value of the vorticity is given by

$$\bar{\omega}(x) = \rho(\vec{x})\cdot 1 = \rho(\vec{x})$$

should also satisfy the constraints:

$$\Gamma(\rho) = \int_\Omega \rho(\vec{x})d\vec{x} = \Gamma_0,$$

$$E(\rho) = \frac{1}{2}\int_\Omega \rho(\vec{x})\psi(\vec{x})d\vec{x} = E_0$$

where

$$\Delta\psi(\vec{x}) = -\rho(\vec{x}).$$

For example, suppose that we have a so-called shear layer as shown in figure 3 as

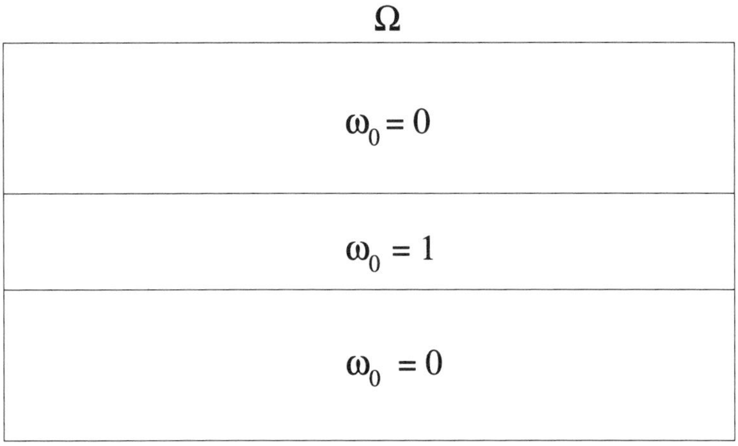

Ω

$\omega_0 = 0$

$\omega_0 = 1$

$\omega_0 = 0$

FIGURE 3. This represents initial data given by a shear layer.

our initial flow $\omega_0(\vec{x})$ and we wish to find the most probable solution which we will guess to be the long time solution. We first compute Γ_0 and E_0 associated with figure 3. Given Γ_0 and E_0 from above, we maximize

$$S(\rho) = -\int_\Omega \rho \log(\rho) + (1-\rho)\log(1-\rho) d\vec{x}$$

subject to

$$\Gamma(\rho) = \int_\Omega \rho d\vec{x} = \Gamma_0$$

and

$$E(\rho) = \frac{1}{2}\int_\Omega \rho\psi d\vec{x} = E_0$$

where

$$\Delta\psi(\vec{x}) = -\rho(\vec{x}).$$

There exists α and β by the Lagrange multiplier rule such that

$$S'(\rho) = \alpha\Gamma'(\rho) + \beta E'(\rho)$$

where in general $F'(\rho)$ denotes a functional derivative defined by

$$F(\rho + \delta\rho) = F(\rho) + < F'(\rho), \delta\rho > + o(\| \delta\rho \|).$$

For our equation, we have

$$S'(\rho) = -\log \frac{\rho}{1-\rho}$$

$$\Gamma'(\rho) = 1$$

$$E'(\rho) = \psi.$$

This implies

$$\rho = \frac{\exp(-\alpha - \beta\psi)}{1 + \exp(-\alpha - \beta\psi)} = -\Delta\psi. \tag{5.1}$$

It is also worth noting that

$$S''(\rho) = -\frac{1}{\rho(1-\rho)} < 0,$$

i.e., S is strictly concave.

6. Numerical Method

We now present our numerical method. In our numerical method, we compute ρ^{k+1} from ρ^k by solving the subproblem

$$S(\rho) \to \text{max, subject to} \tag{6.1}$$

$$\Gamma(\rho) = \Gamma_0, \tag{6.2}$$

$$E(\rho^k) + \int_\Omega E'(\rho^k)(\rho - \rho^k)d\vec{x} \geq E_0. \tag{6.3}$$

S is strictly concave and all the constraints are linear, therefore, the iteration produces a well-defined sequence then if ρ^0 satisfies $\Gamma(\rho^0) = \Gamma_0$ and $E(\rho^0) \geq 0$. The solution ρ^{k+1} satisfies

$$S'(\rho^{k+1}) = \alpha^{k+1}\Gamma'(\rho^{k+1}) + \beta^{k+1}E'(\rho^k)$$

$$\beta^{k+1} \leq 0,$$

$$\beta^{k+1}[E(\rho^k) + < E'(\rho^k), \rho^{k+1} - \rho^k > -E_0] = 0.$$

These are the Kuhn-Tucker conditions, a generalization of the Lagrange multiplier rule for inequalities([**1**],[**7**]). The above conditions imply that if $\beta^{k+1} \neq 0$, the inequality constraint holds with equality. The notation $< \cdot, \cdot >$ denotes the L^2 pairing. In a similar way as (5.1), one can solve for ρ^{k+1} to obtain,

$$\rho^{k+1} = \frac{\exp(-\alpha^{k+1} - \beta^{k+1}\psi^k)}{1 + \exp(-\alpha^{k+1} - \beta^{k+1}\psi^k)} \tag{6.4}$$

The right hand side contains two unknowns α^{k+1} and β^{k+1}. These can be solved uniquely by firstly integrating equation (6.4) and forcing it to satisfy the circulation constraint, equation (6.2) and secondly integrating equation (6.4) and forcing it to satisfy the linear energy constraint (6.3) with equality. This gives 2 nonlinear equations in 2 unknowns at each step with a unique solution. These equations are given by

$$\int_\Omega \rho^{k+1}(\vec{x})d\vec{x} = \Gamma_0 \tag{6.5}$$

$$\int_\Omega \rho^{k+1}(\vec{x})\psi^k(\vec{x})d\vec{x} = E_0 + E^k \tag{6.6}$$

This is a very inexpensive method. We solve the above system by a damped Newton's method. We calculate the integrals in (6.5) and (6.6) numerically using bicubic splines.

7. Dual Problem

The linearized problem (6.1), (6.2) and (6.3) which we solve at each step involves maximizing a functional defined at a huge number of points. In terms of the Lagrange multiplier rule it can be interpreted as follows. Let

$$G(\rho,\alpha,\beta) = S(\rho) - \alpha(\Gamma(\rho)-\Gamma_0) - \beta(E(\rho^k) + < E'(\rho^k), \rho - \rho^k > - E_0).$$

In solving the linearized problem, we find α^{k+1}, β^{k+1} and ρ^{k+1} such that

$$G(\rho^{k+1},\alpha^{k+1},\beta^{k+1}) \geq G(\rho,\alpha,\beta)$$

for every ρ in our admissible class and for all α and $\beta \leq 0$.

There is an associated dual problem which is equivalent to the above problem but results in solving a small system at each step. Minimize $\phi(\alpha,\beta)$ where

$$\phi(\alpha,\beta) = \alpha\Gamma_0 + \beta(E_0 + E(\rho^k)) + \int_\Omega \log(1+\exp(\alpha+\beta\psi^k))d\vec{x} \tag{7.1}$$

with $\alpha \in R$ and $\beta \leq 0$. Minimizing (7.1) is equivalent to solving (6.5) and (6.6).

8. Convergence of the Numerical Method for $\beta < 0$

In section 4, we show that S is strictly concave. This leads to

$$S(\rho+\delta\rho) \leq S(\rho) + < S'(\rho), \delta\rho > -2 \parallel \delta\rho \parallel^2,$$

and using that E is convex, we have

$$E(\rho+\delta\rho) \geq E(\rho) + < E'(\rho), \delta\rho > .$$

By the convexity of E and the inequality constraint (6.3), we have

$$E(\rho^k) \geq E(\rho^{k-1}) + < E'(\rho^{k-1}), \rho^k - \rho^{k-1} > \; \geq E_0.$$

With a little manipulation, we arrive at the principle inequality for our convergence proof,

$$S(\rho^{k+1}) - S(\rho^k) \geq 2 \parallel \rho^{k+1} - \rho^k \parallel^2 + \beta^{k+1}[E_0 - E(\rho^k)]. \tag{8.1}$$

We see that if $\beta < 0$, the entropy increases along the iterative sequence and converges to some limit S^* as $k \to \infty$. We can then conclude that $E(\rho^k) \to E_0$ and that $\rho^{k+1} - \rho^k \to 0$ in L^2. This does not imply convergence of ρ^k which one would not expect due to the nonuniqueness of critical points. However, it is shown in [**8**] that the minimum distance between ρ^k and the set of critical points tends to zero as $k \to \infty$. The algorithm then converges for $\beta < 0$ only. Below, we give a new algorithm which works for positive β along with an alternative algorithm for negative β.

9. Other Algorithms

Another algorithm which we propose for negative β again is based on expanding the entropy to linear terms only and satisfying the energy constraint exactly at each iteration. The solution ρ^{k+1} is given by solving the subproblem,

$$\tilde{S}(\rho) \;\to\; \max \quad \text{subject to}$$

$$\Gamma(\rho) = \Gamma_0, \tag{9.1}$$

$$E(\rho) = E_0 \tag{9.2}$$

where

$$\tilde{S}(\rho) = \int_D S(\rho^k) + S'(\rho^k)(\rho - \rho^k) d\vec{x}$$

We have linearized the entropy above. Using the calculus of variations, we have the variational equation

$$S'(\rho^k) = \beta^{k+1}\psi^{k+1} + \alpha^{k+1}.$$

ρ^{k+1} is obtained uniquely by solving a linear system at each step. We have immediately,

$$\psi^{k+1} = \frac{S'(\rho^k)}{\beta^{k+1}} - \alpha^{k+1}$$

and

$$\rho^{k+1} = -\Delta(\frac{S'(\rho^k)}{\beta^{k+1}})$$

The constant β^{k+1} is then determined by enforcing the circulation constraint(9.1) giving ρ^{k+1}. The multiplier α^{k+1} and ψ^{k+1} are determined by enforcing the energy constraint(9.2). The convergence proof is similar to those in the above algorithm but with a few modifications. Global convergence for $\beta \leq 0$ follows easily from the previous algorithm. Stability must be investigated however for this algorithm.

The algorithms presented above can be proven to work only if $\beta < 0$. We have recently developed an algorithm which can be applied to problems which have solutions corresponding to positive β. The principal inequality in the convergence proofs above is given by (8.1). If $\beta^{k+1} > 0$, then the proof will fail as outlined above unless $E(\rho^k) - E_0 \leq 0$. We can achieve this by solving our linearized maximum entropy problem with our previous iterate ρ^k to get an intermediate solution $\tilde{\rho}^{k+1}$ and obtain the corresponding streamfunction $\tilde{\psi}^{k+1}$. We then solve the following problem for ρ^{k+1}.

$$S(\rho) \rightarrow \max \quad \text{subject to}$$

$$\Gamma(\rho) = \Gamma_0,$$

$$E(\rho^k) + \int_D \tilde{\psi}^{k+1}(\rho - \rho^k)dx \leq E_0.$$

Two problems must be resolved at each step, however, it appears that this code exhibits a faster rate of convergence than the methods presented before. Another method for this problem is given also by [**6**].

References

[1] A.D. Ioffe and V.M. Tihomirov. *Theory of extremal problems.* Elsevier North-Holland, 1979.
[2] J. Miller. Statistical mechanics of Euler equations in two dimensions. *Phys. Rev. Lett.*, 65:2137, 1990.
[3] J. Miller, P. Weichman, and M.C. Cross. Statistical mechanics, Euler's equations, and Jupiter's red spot. *Phys. Rev. A*, 45:2328, 1992.
[4] R. Robert. A maximum-entropy principle for two-dimensional perfect fluid dynamics. *J. Stat. Phys.*, 65:531, 1991.
[5] R. Robert and J. Sommeria. Statistical equilibrium states for two-dimensional flows. *J. Fluid Mech.*, 229:291, 1991.
[6] R. Robert and J. Sommeria. Relaxation towards a statistical equilibrium state in two-dimensional perfect fluid dynamics. *Phys. Rev. Lett.*, 69:2776, 1992.

[7] R.T. Rockafellar. *Convex analysis.* Princeton Univ. Press, Princeton, N.J., 1970.
[8] B. Turkington and N. Whitaker. Statistical equilibrium computations of coherent structures in turbulent shear layers. *SIAM J. Comput.*, 17:1414, 1996.
[9] N. Whitaker and B. Turkington. Maximum entropy states for rotating vortex patches. *Phys. Fluids A*, 6:3963, 1994.

DEPARTMENT OF MATHEMATICS AND STATISTICS, UNIVERSITY OF MASSACHUSETTS, AMHERST, MA 01002

E-mail address: whitaker@math.umass.edu

DIMACS Series in Discrete Mathematics
and Theoretical Computer Science
Volume **34**, 1997

Hydrodynamic Stability, Differential Operators and Spectral Theory

Isom H. Herron

"For all that has been - Thanks!
To all that shall be - Yes!" - Dag Hammarskjöld

1. Introduction

The April, 1996 SIAM NEWS had the following headline for an article by Barry A. Cipra: "Superpipe: An Experimental Gauge for Computational Fluid Dynamics". It was accompanied by a picture of a 100 foot long apparatus. The article went on to describe the theoretical and experimental effort being carried out at Princeton University. To those of us who have worked in the field for years, this is truly sweet justification after hearing comments of others that the field was "dead". It is truly alive!

The subject of fluid mechanics has been of interest to scientists and mathematicians for centuries. Over the last one hundred fifty years, it has been rationalized and codified, beginning with the work of Navier and Stokes in the last century. Continuum mechanics has been developed more recently, but during its development fluid mechanics was invariably perceived as an important subfield. The mathematical theory of the Navier-Stokes equations of fluid mechanics has grown steadily since the 1930's. Much of the best of modern mathematics has been applied to the study of these equations [Galdi, 1995], [Doering and Gibbon, 1995]. The area of incompressible fluid mechanics, or hydrodynamics, takes as its basis the Navier-Stokes equations.

The goal of this presentation is to treat some classical, exact, steady, two-dimensional solutions of the Navier-Stokes equations, having the property of constant vorticity. The stability of these solutions to time-dependent, two-dimensional perturbations leads to the spectral problem of hydrodynamic

---

1991 Mathematics Subject Classification. Primary 76D05, 76E05; Secondary 35Q30, 47B25.

The author was supported by a grant from the Office of Naval Research, Applied Analysis Program.

stability.

The unifying property of constant vorticity permits the stability problems to be treated in a single formulation. The organization of the paper is as follows. In the next section, the Navier-Stokes equations are introduced. In the succeeding two sections, exact solutions for special geometries are presented as an illustration, and the notion of these solutions is extended to a more general geometry. The stability problem is defined in section 5, and in section 6, a comparison problem is treated. Then, in section 7, a stability proof is sketched for the flows described in section 4, which is believed to be a new result. The last section of the paper gives possible further directions for this area of research.

2. Equations of Motion

The Navier-Stokes equations with velocity $\mathbf{u} = (u, v, w) = \mathbf{u}(\mathbf{x}, t)$, $\mathbf{x} = (x_1, x_2, x_3)$

$$\frac{\partial \mathbf{u}}{\partial t} + \mathbf{u} \cdot \nabla \mathbf{u} = \frac{-\nabla p}{\rho} + \nu \nabla^2 \mathbf{u},$$

reflect Newton's 2nd law. Sometimes these are called the momentum equations. Density ρ, viscosity ν, pressure p are the parameters in equation, which may be variables in general. Conservation of mass gives the continuity equation

$$\frac{\partial \rho}{\partial t} + \nabla \cdot (\rho \mathbf{u}) = 0.$$

When $\rho = \text{const}$, $\nabla \cdot \mathbf{u} = 0$, so the velocity field is solenoidal. Appropriate to the partial differential equations are boundary conditions: no slip at an impermeable surface. Another important dynamic variable is vorticity $\omega = \nabla \times \mathbf{u}$. Taking the curl of the momentum equations eliminates pressure:

$$\frac{\partial \omega}{\partial t} + \mathbf{u} \cdot \nabla \omega - \omega \cdot \nabla \mathbf{u} = \nu \nabla^2 \omega.$$

However, no boundary conditions are available on vorticity in general.

3. Exact Solutions

(i) Channel flow [Hunt, 1964]. The set-up is an infinitely long two dimensional channel whose parallel walls are separated by a distance h. Suppose one wall moves with a constant velocity U_0 in the x-direction. A steady flow is sought with velocity components (u, v, w), where $v = w = 0$. Continuity is automatically satisfied when $u = u(y)$. The pressure is assumed to depend on x only. The only nonzero component of the momentum equations is

$$0 = -\frac{1}{\rho}\frac{dp}{dx} + \nu \frac{\partial^2 u}{\partial y^2}.$$

Take $u(0) = 0, u(h) = U_0$ giving

$$\bar{u}(y) = \frac{U_0 y}{h} - \frac{dp}{dx}\frac{yh}{\rho\nu}\left(1 - \frac{y}{h}\right).$$

This is called combined plane Poiseuille-Couette flow. If $dp/dx = 0$, then plane Couette flow results which is sometimes called a uniform shear flow. It has the property that its vorticity is constant,

$$\bar{\omega} = -(U_0/h)\hat{k}.$$

See Figure (i).

(ii) Couette Flow [Hunt, 1964]. Steady laminar flow between concentric rotating cylinders is described in cylindrical coordinates (r, θ, z). The velocity components are (u, v, w), with $u = w = 0$. By continuity v is a function of r only. The momentum equations reduce to

$$\frac{v^2}{r} = \frac{1}{\rho}\frac{dp}{dr} : \text{radial direction},$$

$$\frac{d^2 v}{dr^2} + \frac{d}{dr}\left(\frac{v}{r}\right) = 0 : \text{circumferential direction}.$$

Boundary conditions are again no-slip

$$v(a) = a\Omega_1, \; v(b) = b\Omega_2.$$

To solve, the circumferential equation is integrated (twice), introducing two arbitrary constants, which are determined by the boundary conditions. The pressure is given by the equation in the radial direction. This flow also has the constant vorticity property. Thus

$$\bar{\omega} = 2(\frac{\Omega_1 a^2 - \Omega_2 b^2}{a^2 - b^2})\,\hat{k}.$$

See Figure(ii).

Familiar 2-D flows with constant vorticity

1. Plane Couette Flow

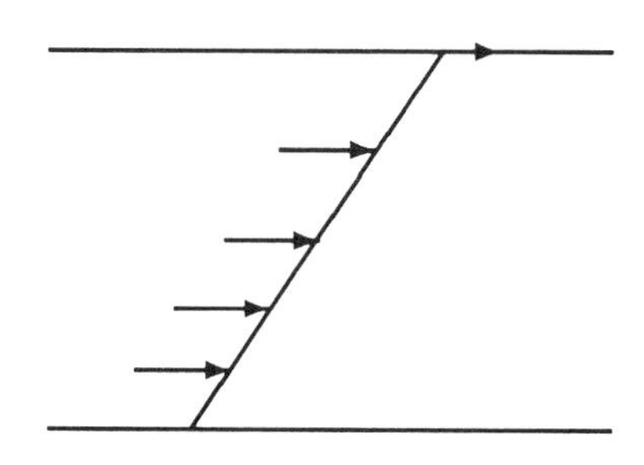

Fig (i)

$$\bar{u} = \frac{U_0 y}{h} \qquad \text{"Simple Shear": Parallel Streamlines}$$

2. Couette Flow Between Cylinders

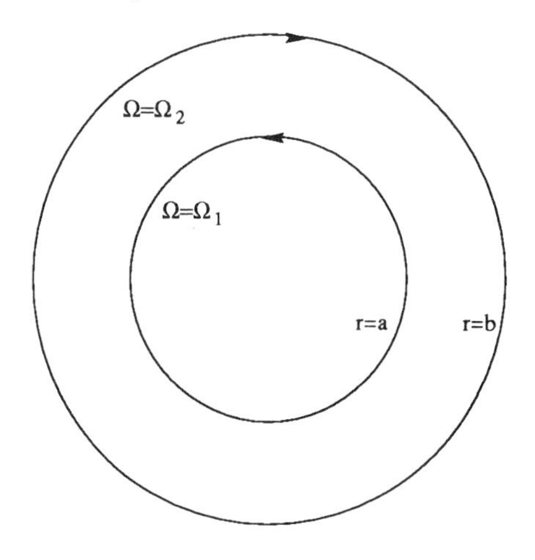

Fig. (ii)

$$\bar{v}(r) = \frac{1}{r}\left(\frac{\Omega_1 - \Omega_2}{a^{-2} - b^{-2}}\right) + r\left(\frac{\Omega_1 a^2 - \Omega_2 b^2}{a^2 - b^2}\right) : \text{Circular Streamlines}$$

4. General Considerations

Consider now a two-dimensional viscous flow (of magnitude U_0) on a bounded region (of length scale L) called $D \subset R^2$, whose boundary is given by ∂D, sufficiently smooth that Green's theorem in the plane is known to hold. The physical possibility that such a flow is useful is given in the classic reference [Batchelor,1956].

Exact solutions $\mathbf{u} = (u, v, 0)$ to the 2-D vorticity equation

$$\frac{\partial \omega}{\partial t} + \mathbf{u} \cdot \nabla \omega = \frac{1}{R} \Delta \omega,$$

are sought where $R = \dfrac{U_0 L}{\nu}$ is the Reynolds number and $\omega = \dfrac{\partial v}{\partial x} - \dfrac{\partial u}{\partial y}$ is the only non-zero component of the vorticity.
Introduce a stream function ψ

$$u = \frac{\partial \psi}{\partial y}, \quad v = -\frac{\partial \psi}{\partial x}.$$

Hence $\omega = -\Delta \psi$ and

$$\frac{\partial}{\partial t} \Delta \psi - \frac{\partial(\psi, \Delta \psi)}{\partial(x, y)} = \frac{1}{R} \Delta^2 \psi.$$

Steady flows with vorticity constant satisfy $-\Delta \psi_0 = \omega_0$ constant in D, Poisson's equation. No-slip boundary conditions are $(u, v) = (u_0, v_0)$ on the boundary ∂D.

5. Stability Problem

The basic steady solution is perturbed written

$$\psi(x, y, t) = \psi_0(x, y) + \hat{\psi}(x, y, t),$$

giving

$$\frac{\partial}{\partial t} \Delta \hat{\psi} - \frac{\partial(\psi_0, \Delta \hat{\psi})}{\partial(x, y)} - \frac{\partial(\hat{\psi}, \Delta \hat{\psi})}{\partial(x, y)} = \frac{1}{R} \Delta^2 \hat{\psi},$$

where $\hat{\psi} = 0$ and $\dfrac{\partial \hat{\psi}}{\partial n} = 0$ on ∂D.

Linearize and set

$$\hat{\psi} = \phi(x, y) e^{-\sigma t},$$

resulting in an elliptic boundary value problem of 4th order

$$-\sigma\Delta\phi - \frac{\partial\psi_0\partial}{\partial x\partial y}\Delta\phi + \frac{\partial\psi_0\partial}{\partial y\partial x}\Delta\phi = \frac{1}{R}\Delta^2\phi, \qquad [OS]$$

where $\phi = \frac{\partial\phi}{\partial n} = 0$ on ∂D.

Make the identification $\zeta = -\Delta\phi$ on D to obtain

$$-\sigma\zeta - J(\psi_0, \zeta) = \frac{1}{R}\Delta\zeta. \qquad [OS]$$

There are no boundary conditions on ζ, and $J(\psi_0, \zeta) \equiv \frac{\partial\psi_0\partial\zeta}{\partial x\partial y} - \frac{\partial\psi_0\partial\zeta}{\partial y\partial x}$.

The equation $[OS]$ is called the generalized Orr-Sommerfeld equation. If for a given Reynolds number R, all eigenvalues satisfy

$$Re(\sigma) > 0,$$

the flow is said to be linearly stable. If, for some R, there occurs at least one eigenvalue σ satisfying

$$Re(\sigma) < 0,$$

the flow is said to be unstable. The usual scenario is that for R sufficiently small the flow is linearly stable. Then as R is increased beyond a certain critical value of R called R_c, the flow is unstable [Joseph, 1976], [Straughan, 1992]. However, for the two simple flows with constant vorticity, figs (i), (ii), no critical Reynolds number exists for instability to two-dimensional or plane disturbarnces. The purpose of the rest of this article is to show that this is to be expected for these and any two-dimensional flow with constant vorticity [Ali and Herron, 1996], [Herron, 1992].

6. Comparison Problem

Consider a reduced form of the linearized vorticity equation $[S]$:

$$\begin{aligned} -\lambda\Delta\phi &= \Delta^2\,\phi \quad \text{on} \quad \mathrm{D}, \\ \phi = \frac{\partial\phi}{\partial n} &= 0 \quad \text{on} \quad \partial\mathrm{D}, \end{aligned}$$

giving eigenvalues $\{\lambda_j\}$, and eigenfunctions $\{\phi_j\}$. This is analogous to the buckling of a plate. One selfadjoint realization implies the normalizations

$$\langle\phi_m, \Delta\phi_j\rangle = \delta_{mj}$$

where

$$\begin{aligned} -\langle \phi_m, \Delta\phi_j \rangle &= -\iint_D \phi_m \Delta\phi_j dA \\ &= \iint_D \nabla\phi_m \cdot \nabla\phi_j dA. \end{aligned}$$

Furthermore

$$\left.\begin{aligned} \iint_D \Delta^2\phi_m \Delta\phi_j dA &= -\lambda_m \iint_D \Delta\phi_m \Delta\phi_j dA \\ \iint_D \Delta^2\phi_j \Delta\phi_m dA &= -\lambda_j \iint_D \Delta\phi_j \Delta\phi_m dA \end{aligned}\right\} \Rightarrow$$

$$(\lambda_j - \lambda_m) \iint_D \Delta\phi_j \Delta\phi_m dA =$$

$$= \iint_D (\Delta^2\phi_m \Delta\phi_j - \Delta^2\phi_j \Delta\phi_m) dA.$$

By Green's identity

$$\iint_D (\Delta^2\phi_j \Delta\phi_m - \Delta^2\phi_m \Delta\phi_j) dA =$$

$$= \oint_{\partial D} \left(\Delta\phi_m \frac{\partial}{\partial n} \Delta\phi_j - \Delta\phi_j \frac{\partial}{\partial n} \Delta\phi_m \right) ds.$$

Why this equals 0 follows from the results of the following.

Lemma 1:

Let $F(x, y)$ be a harmonic function in D. Then, for ϕ satisfying $\phi = \frac{\partial\phi}{\partial n} = 0$ on ∂D,

(i) $\iint_D \Delta\phi F dA = 0$,

and for ϕ satisfying the above and $\Delta^2\phi = -\lambda\Delta\phi$,

(ii) $\oint_{\partial D} \left(F \frac{\partial}{\partial n} \phi - \frac{\partial F}{\partial n} \Delta\phi \right) ds = 0$.

Proof:

(i) Use Green's identity.

(ii) Use part (i) and Green's identity. □ A second type of orthogonality is thereby obtained for [*S*]:

$$\langle \Delta\phi_j, \Delta\phi_m \rangle = 0, \quad j \neq m.$$

This permits consideration of another symmetric eigenvalue problem [S′]:

$$-\Delta\zeta = \lambda\zeta \quad \text{in} \quad \mathrm{D}$$

$$\oint_{\partial D}\left(F\frac{\partial\zeta}{\partial n} - \frac{\partial F}{\partial n}\zeta\right)ds = 0,$$

for all F satisfying $\Delta F = 0$ in D. The eigenvalues of the problem [S′] are identical to those of the problem [S], except that $\lambda_0 = 0$ is an additional eigenvalue of [S′] corresponding to the eigenfunctions $\zeta_0 = \{F\}$.

Lemma 2: *Suppose M is a closed symmetric operator in a Hilbert space $\mathcal{H}$, positive bounded below with bound b, and with closed range. Then M has a unique closed adjoint M^*, with closed range [Kato]. Define an operator $\hat{M}$ with domain*

$$\mathrm{dom}\hat{\mathrm{M}} = \{\phi \in \mathrm{dom}\ \mathrm{M}^* | \mathrm{M}^*\phi \perp \mathrm{nul}\mathrm{M}^*\}$$

such that

$$\hat{M}\phi = M^*\phi, \quad \phi \in \mathrm{dom}\ \mathrm{M}^*.$$

Then $\hat{M}$ is a selfadjoint extension of M and

$$\langle \hat{M}\phi, \phi\rangle \geq 0, \quad \phi \in \mathrm{dom}\ \hat{\mathrm{M}}.$$

Note: This extension is the von Neumann extension, which he introduced in 1929.

Lemma 2 is applied by taking $M^* = -\Delta$, with no boundary conditions, while $M = -\Delta$, with both Dirichlet and Neumann boundary conditions. (A little later in 1932 von Neumann, proved that M^*M is a selfadjoint positive semi-definite operator, even when M is not necessarily symmetric bounded below.) Thus, $\hat{M} = -\Delta$, with the integral boundary conditions in [S′]. However, the null space, nul $\hat{M}$ is infinite dimensional.

7. Stability of 2-D Flows with Constant Vorticity

Lemma 3: *If $\zeta = M\phi$ satisfies [OS], then ζ also satisfies* [OS′]

$$\hat{M}\zeta - RJ(\psi_0, \zeta) = \lambda\zeta,$$

where $\lambda = \sigma R$.

This leads to the desired conclusion.

Theorem 1: *Two-dimensional flows with constant vorticity are stable to 2-D disturbances.*

Proof: It follows from Lemma 3 that,

$$< \hat{M}\zeta, \zeta > -R < J(\psi_0, \zeta), \zeta > = \lambda < \zeta, \zeta > .$$

Use a complex inner product, since λ will be complex

$$< J(\psi_0, \zeta), \zeta >= \iint_D \left(\frac{\partial \psi_0}{\partial x}\frac{\partial \zeta}{\partial y} - \frac{\partial \psi_0}{\partial y}\frac{\partial \zeta}{\partial x} \right) \bar{\zeta} dA.$$

Put $\zeta = \zeta_1(x,y) + i\zeta_2(x,y), |\zeta|^2 = \zeta_1^2 + \zeta_2^2$. Then

$$< J(\psi_0, \zeta), \zeta >=$$

$$-\iint_D \left[\frac{\partial}{\partial x}\left(\frac{|\zeta|^2}{2}\frac{\partial \psi_0}{\partial y} \right) - \frac{\partial}{\partial y}\left(\frac{|\zeta|^2}{2}\frac{\partial \psi_0}{\partial x} \right) \right] dA$$

$$+ i \iint_D [\zeta_1 J(\psi_0, \zeta_2) - \zeta_2 J(\psi_0, \zeta_1)] dA$$

which is pure imaginary.

Thus, $Re(\lambda) =< \hat{M}\zeta, \zeta > / \|\zeta\|^2 > 0$. □

Theorem 1 depends on the proof of Lemma 3.

Sketch of Proof of Lemma 3: The following representation is due to von Neumann: dom$\hat{M}$ = dom$M \oplus$ nulM* $\equiv \mathcal{D}_{\mathrm{M}} \oplus \mathcal{N}_0$.

Consider solutions of the eigenvalue problem

$$\hat{M}Z + RJ(\psi_0, Z) = \mu Z, \qquad Z \in \text{dom } \hat{M}.$$

This is an operator perturbation of [S′]. Thus it has a complete set of eigenfunctions in $\mathcal{H}$ [DiPrima and Habetler, 1969].

Write $Z_k = h_k + f_k$, $\quad k = 1, 2, \ldots$, where $h_k \in \mathcal{D}_M, f_k \in \mathcal{N}_0$.

It followis that

$$\hat{M}(h_k + f_k) + RJ(\psi_0, h_k + f_k) = \mu_k(h_k + f_k),$$

or

$$Mh_k + RJ(\psi_0, h_k) + RJ(\psi_0, f_k) = \mu_k h_k + \mu_k f_k.$$

Apply M^*

$$M^*Mh_k + RM^*(J(\psi_0, h_k)) + RM^*(J(\psi_0, f_k)) = \mu_k M h_k.$$

The Fredholm Alternative gives that solutions exist only if

$$R\langle M^*(J(\psi_0, f_k), \phi_k\rangle = 0.$$

That is if $\langle J(\psi_0, f_k), M\phi_k\rangle = 0$, where ϕ_k satisfies $[OS]$

$$M^*M\phi_k - RJ(\psi_0, M\phi_k) = \bar{\mu}_k M\phi_k.$$

Identify $\bar{\lambda}_k = \mu_k$, since R is fixed.

Conclusion:

$$J(\psi_0, M\phi_k) \perp \mathcal{N}_0.$$

This condition is met when (and only when)

$$M^*\zeta_k \perp \mathcal{N}_0,$$

since $\zeta_k \perp \mathcal{N}_0$. Thus if ζ_k satisfies $[OS]$, it also satisfies $[OS']$.□

8. Summary

A proof has been presented that two-dimensional steady viscous flows with constant vorticity are linearly stable to two-dimensional time dependent perturbations. This unifies the two-dimensional stability phenomena of plane Couette flow and Couette flow between rotating cylinders. What is more this general proof, for all constant vorticity flows is new. An important question is whether the use of the von Neumann theory of operators may be successfully applied to other newer problems in fluid dynamics, such as the dimension of attractors, etc [Doering and Gibbon, 1995].

9. References

Ali, H. N. and Herron I. H., "The Two-Dimensional Stability of a Viscous Fluid between Rotating Cylinders," J. Math. Anal. Appl. **203**, 482-490 (1996).

Batchelor, G. K., "On steady laminar flow with closed streamlines at large Reynolds number," J. Fluid Mech. 1, 177-190 (1956).

DiPrima, R. C. and Habetler, G. J., "A Completeness Theorem for Non-selfajoint Eigenvalue Problems in Hydrodynamic Stability," Arch Rat. Mech. Anal. **34**, 218-227 (1969).

Doering, C. R. and Gibbon, J. D., *Applied Analysis of the Navier-Stokes Equations*, Cambridge University Press, Cambridge (1995).

Galdi, G. P., *An Introduction to the Mathematical Theory of the Navier-Stokes Equations, 2 Vols.*, Springer-Verlag, New York (1995).

Herron, I. H., "The Linear Stability of Circular Pipe Flow to Axisymmetric Disturbances," Stab. Appl. Anal. Cont. Media **2**, 293-302 (1992).

Hunt, J. N., *Incompressible Fluid Dynamics*, Longmans, London. (1964).

Joseph, D. D., *The Stability of Fluid Motions, 2 Vols.*, Springer-Verlag, New York (1976).

Kato, T., *Perturbation Theory for Linear Operators (2nd ed).*, Springer-Verlag, New York (1976).

Straughan, B., *The Energy Method, Stability and Nonlinear Convection*, Springer-Verlag, New York (1992).

Department of Mathematical Sciences, Rensselaer Polytechnic Institute, Troy, NY 12180-3590

DIMACS Series in Discrete Mathematics
and Theoretical Computer Science
Volume **34**, 1997

The Role of Selberg's Trace Formula in the Computation of Casimir Energy for Certain Clifford-Klein Space-Times

Floyd L. Williams

ABSTRACT: Physicists have pointed out that the problem of computing the topological Casimir energy E for massless real scalar fields on space-times $S = R \times M$ for various manifolds M is of considerable interest and importance in areas of quantum field theory, quantum cosmology, hadronic physics, and the like. We consider the interesting case where M is a compact space form $\Gamma \setminus X$ modeled on a rank 1 symmetric space X, with Γ a discrete group of isometries of X. In this general setting one has a powerful, specific mathematical tool - the Selberg trace formula. Using this formula we show in general that E has a beautiful expression in terms of the spherical harmonic analysis of X and Selberg's zeta function. Our work extends that of Bytsenko, Goncharov, Zerbini, and others, who focused on the real hyperbolic case $X = SO_1(n,1)/SO(n)$. Specifically, we extend the work of these authors to the other classical rank 1 irreducible symmetric spaces $X = SU(n,1)/U(n)$, $SP(n,1)/(SP(n) \times SP(1))$, and the exceptional space $X = F_{4(-20)}/Spin(9)$.

We also consider implications of the trace formula for non-compact $\Gamma \setminus X$ (for Γ with co-finite volume), in response to remarks of Bytsenko and Goncharov that such quotients should figure in the next step of the investigation of the Casimir effect.

1. Introduction

We shall be working with an irreducible rank 1 symmetric space $X = G/K$ of non-compact type: G will be a connected non-compact simple split rank 1 Lie group with finite center and $K \subset G$ will be a maximal compact subgroup. We choose the following representations of X up to local isomorphism:

1996 *Mathematics Subject Classification*: Primary 22E46, 17B29, 11F72, 11M41. Secondary 43A85, 43A90, 32N15.

Key Words and Phrases: Selberg trace formula, Casimir energy, Clifford-Klein space-time, Harish-Chandra c-function, Minakshisundaram-Pleijel zeta function, Gangolli-Selberg zeta function, scattering matrix, Eisenstein series.

$$X = \begin{bmatrix} SO_1(n,1)/SO(n) & (i) \\ SU(n,1)/U(n) & (ii) \\ SP(n,1)/(SP(n)\times SP(1)) & (iii) \\ \text{or} & \\ F_{4(-20)}/Spin(9) & (iv) \end{bmatrix} \tag{1.1}$$

where $n \geq 2$, and $F_{4(-20)}$ is the unique real form of F_4 (with Dynkin diagram O–O=O–O) for which the character $\delta \overset{def}{=} \dim X - \dim K$ assumes the value -20 [22]. Let $\Gamma \subset G$ be a discrete subgroup. For now we assume $\Gamma \setminus G$ is compact and Γ is torsion free; later we allow $\Gamma \setminus G$ to be non-compact but of finite G-invariant volume. Thus $M_\Gamma \overset{def}{=} \Gamma \setminus X$ is a compact manifold. $X_\Gamma \overset{def}{=} R \times M_\Gamma$ (for the field of real numbers R) is a space-time of Clifford-Klein type modeled on X.

Recently, some physicists have focused with great interest on the case $X_\Gamma = X_\Gamma(n) \overset{def}{=} R\times(\Gamma \setminus SO_1(n,1)/SO(n))$ (which corresponds to case (i) in (1.1)) in connection with the problem of evaluating the Casimir energy (or vacuum effect) for massless scalar fields (or spinor fields) on $X_\Gamma(n)$ [5], [6], [7], [8], [9], [11]. The mathematical problem involved here can be described (for general X_Γ) as follows. Let χ be a finite-dimensional unitary representation of Γ on a vector space V_χ. Regarding χ as a trivial automorphy factor $\Gamma \times X \to GL(V_\chi)$, $(\gamma,\chi)\to\chi(\gamma)$, we have the corresponding vector bundle $V_\chi \overset{def}{=} \Gamma \setminus (X \times V_\chi) \to \Gamma \setminus X = M_\Gamma$ with fiber V_χ where Γ acts on $X \times V_\chi$ via $\chi : \gamma . (x , v) = (\gamma.x, \chi(\gamma) v)$ for $(\gamma,x,v) \in \Gamma \times X \times V_\chi$. The Laplacian on X being G - invariant (and hence Γ - invariant) projects to a differential operator Δ_Γ on smooth sections of V_χ. Let $\{\lambda_j=\lambda_j(\chi)\}_{j=0}^{\infty}$ be the corresponding set of eigenvalues of $-\Delta_\Gamma$ and let $n_j(\chi)$ denote the multiplicity of λ_j. Because of the minus sign preceding Δ_Γ the eigenvalues can be labeled to satisfy $0=\lambda_0<\lambda_1<\lambda_2<\ldots$; $\lim_{j\to\infty} \lambda_j = \infty$. Assume for example that $d \overset{def}{=} \dim X$ is even, which is the case for all X in (1.1) except for case (i) where n is odd. Then by formal considerations (canonical quantization of scalar fields [1], [2], [3], [17]) physicists argue that the desired expression for the Casimir energy $E_v(\Gamma,\chi)$ is given by

$$E_v(\Gamma,\chi) = \tfrac{1}{2} \sum_j n_j(\chi)\lambda_j^{1/2} \tag{1.2}$$

We shall also write $E_v(\chi)$ for $E_v(\Gamma,\chi)$ and also call $E_v(\chi)$ the vacuum energy, which is the reason for the subscript v. The series in (1.2) has no chance of converging. The mathematical problem is therefore to assign a meaning to it by some regularization process. We employ zeta regularization [6], [10], [15], [20]. Namely let

$$D_\Gamma(s;\chi) = \sum_{j=1}^{\infty} \frac{n_j(\chi)}{\lambda_j^s} \tag{1.3}$$

be the Minakshisundaram-Pleijel (M-P) spectral zeta function of $-\Delta_\Gamma$, which is a holomorphic function of s on Res > d/2 [28]. Although the series (1.3) do not converge for $s= -\frac{1}{2}$ one has formally $E_v(\chi) = \frac{1}{2}D_\Gamma(-\frac{1}{2};\chi)$. On the other hand the author, using the Selberg trace formula, has worked out an explicit meromorphic continuation of $D_\Gamma(s;\chi)$ to the full complex plane **C** in terms of the spherical harmonic analysis of X (for all the rank 1 symmetric spaces X in (1.1)); see [36]; also see [41] and see [30] for the initial special case $X = SO_1(2,1) / SO(2) = SL(2,R) / SO(2)$. In particular the meromorphic continuation, which we also denote by $D_\Gamma(s;\chi)$, is holomorphic at $s = -\frac{1}{2}$ for dim X even. We can therefore consider $E_v(\chi)$ as well-defined by assigning it the value $\frac{1}{2}D_\Gamma(-\frac{1}{2}\,;\chi)$. On the other hand, in [42] we show in general (even for dim X odd) that on the domain Res < 0 there is a beautiful expression of the M-P zeta function $D_\Gamma(s;\chi)$ in terms of the logarithmic derivative $\Psi_\Gamma(s;\chi)$ of the Selberg zeta function $Z_\Gamma(s;\chi)$ attached to the data (G, K, Γ,χ) [18] (see Theorem 3.5 below) - an expression first obtained by the physicists in the case $X = SO_1(n,1)/SO(n)$ [5], [6], [7], [8], [9]. In particular we can express $E_v(\chi)$ neatly in terms of $\Psi_\Gamma(s;\chi)$ (and the harmonic analysis of X) and discover that for the trivial line bundle $V_1 \to M_\Gamma$ the topological component of $E_v(1)$ is always negative.

We should point out that in application the bundles $V_\chi \to M_\Gamma$ are taken to be line bundles in fact corresponding to characters χ of Γ mod 2. Namely, according to [23], the topologically inequivalent configurations of real scalar fields correspond univalently to elements of the first cohomology group $H^1(X_\Gamma, Z_2)$ of the space-time X_Γ with coefficients in Z_2. This cohomology on the other hand is given by $\mathrm{Hom}(\Pi_1(X_\Gamma),Z_2)$ where $\Pi_1(X_\Gamma)$ is the fundamental group of X_Γ, and thus is given by characters of Γ mod 2 since $\Pi_1(X_\Gamma) = \Pi_1(R) \times \Pi_1(M_\Gamma) = \Pi_1(M_\Gamma) = \Gamma$.

When dim X is odd, i.e. $G = SO_1(n,1)$ with n odd, $D_\Gamma(s;\chi)$ has a simple pole at $s = -\frac{1}{2}$. In this case $E_v(\chi)$ can still be defined by a more general regularization formula involving the (necessarily non-vanishing) heat equation coefficient K_{n+1} of X [5], [6], [9].

The general form of the trace formula is required to construct the function $\Psi_\Gamma(s;\chi)$[18]. We use also the special formula that results when one plugs in the Harish-Chandra-Schwartz test function h_t, $t > 0$, where up to scaling h_t is the fundamental solution of the heat equation on X - a generalized Jacobi inversion formula for a suitable ϑ-function $\vartheta_\Gamma(t)$. For by taking the

Mellin transform $\int_0^\infty \ldots t^{s-1}dt$ of both sides of the latter formula we capture $D_\Gamma(s;\chi)$ (up to a Γ-factor) on the one hand, and its meromorphic continuation (after some work) on the other hand. Compare equations (4.6), (4.8) and (4.13) in section 4 below.

2. Harish-Chandra's Plancherel density

The spherical harmonic analysis on $X = G / K$ is described in terms of Harish-Chandra's c-function [12], [13], [22], which we may regard as a meromorphic function c(z) on C, since the rank of X is 1. The Plancherel density $|c(r)|^{-2}$ on R also shows up in connection with the expression of the contribution of the identity element of Γ to the trace formula. We make use of Miatello's computation of this density [26],[27]. Namely, for suitable normalization of Haar measures (see [36],[38], or [41] for specifics), $|c(r)|^{-2}$ is given as follows, where C_G is a constant depending on G.

$$|c(r)|^{-2} = \begin{cases} C_G\pi r P(r)\tanh \pi r & \text{for } G=SO_1(2n,1) \\ C_G\pi P(r) & \text{for } G=SO_1(2n+1,1) \\ C_G\pi r P(r)\tanh \frac{\pi}{2} r & \text{for } G=SP(n,1) \text{ or } F_{4(-20)} \\ C_G\pi r P(r)\begin{bmatrix} \tanh\frac{\pi}{2}r \\ \text{or} \\ \coth\frac{\pi}{2}r \end{bmatrix} & \begin{matrix}\text{for } G=SU(n,1) \text{ where} \\ \text{the cotangent choice is} \\ \text{made for n even}\end{matrix} \end{cases} \tag{2.1}$$

where P(r) is an even polynomial of degree d-2 for $G \neq SO(2n+1,1)$, and of degree d-1 $= 2n$ for $G = SO_1(2n+1,1)$ given as follows:

$$P(r) = \prod_{j=2}^{n}[r^2 + (n-j+\tfrac{1}{2})^2] = \prod_{j=0}^{n-2}[r^2 + \frac{(2j+1)^2}{4}] \qquad \text{for } G = SO_1(2n,1),\ n \geq 1$$

$$P(r) = \prod_{j=1}^{n}[r^2 + (n-j)^2] = \prod_{j=0}^{n-1}[r^2 + j^2] \qquad \text{for } G = SO_1(2n+1,1),\ n \geq 1$$

$$P(r) = [\frac{r^2+1}{4}]\prod_{j=3}^{n+1}[\frac{r^2}{4} + (n-j+\tfrac{3}{2})^2]\,[\frac{r^2}{4} + (n-j+\tfrac{5}{2})^2] \qquad \text{for } G = SP(n,1), n \geq 2$$

$$\tag{2.2}$$

$$P(r) = [\frac{r^2+1}{4}]^2[\frac{r^2}{4} + (\tfrac{3}{2})^2]^2[\frac{r^2}{4} + (\tfrac{5}{2})^2][\frac{r^2}{4} + (\tfrac{7}{2})^2][\frac{r^2}{4} + (\tfrac{9}{2})^2] \qquad \text{for } G = F_{4(-20)}, \text{ and}$$

$$P(r) = \prod_{j=1}^{n-1}[\frac{r^2}{4} + \frac{(n-2j)^2}{4}] \qquad \text{for } G = SU(n,1),\ n \geq 2$$

Denote by a_{2j} the coefficients of the polynomials $P(r)$ in (2.2):

$$P(r) = \sum_{j=0}^{(d/2)-1} a_{2j} r^{2j} \qquad \text{for } G \neq SO_1(2n+1,1)$$

$$P(r) = \sum_{j=0}^{n} a_{2j} r^{2j} \qquad \text{for } G = SO_1(2n+1,1) \tag{2.3}$$

where again $d = \dim X$. Note that

$$d = n,\ 2n,\ 4n,\ 16 \tag{2.4}$$

in cases (i), (ii), (iii), (iv) respectively in (1.1). We shall need the ½ sum ρ_0 of the positive restricted (real) roots of G (with multiplicity) with respect to a nilpotent factor N of G in an Iwasawa decomposition $G = KAN$. ρ_0 is given as follows

$$\rho_0 = \frac{n-1}{2},\ n,\ 2n+1,\ 11 \tag{2.5}$$

in cases (i), (ii), (iii), (iv) respectively in (1.1).

Given $a, \delta > 0$ we define entire functions K_n, L_n of s for $n = 0, 1, 2, 3, \ldots$ by

$$K_n(s;\delta,a) = \int_R \frac{r^{2n}\operatorname{sech}^2 ar}{(\delta + r^2)^s}\, dr$$

$$L_n(s;\delta,a) = \int_R \frac{r^{2n+1}(\operatorname{csch} ar)\operatorname{sech} ar}{(\delta + r^2)^s}\, dr \tag{2.6}$$

Using the functions K_n, L_n, the constants C_G in (2.1), and the coefficients a_{2j} in (2.3) we construct a meromorphic function $I(s\,;b)$ of s, where $b \geq 0$ is fixed, as follows. For $G \neq SO_1(m,1)$, $SU(p,1)$ with m odd and p even

$$I(s;b) \stackrel{\text{def}}{=} \frac{a(G)}{2} C_G \pi \sum_{j=0}^{(d/2)-1} a_{2j}\, j! \sum_{\ell=0}^{j} \frac{K_{j-\ell}(s-\ell-1; b+\rho_0^2, a(G))}{(j-\ell)!(s-1)(s-2)\ldots(s-(\ell+1))}$$

$$\text{where } a(G) = \begin{cases} \pi & \text{for } G = SO_1(2n,1) \\ \dfrac{\pi}{2} & \text{for } G = SU(p,1) \text{ with p odd} \\ & \text{or } \ G = SP(m,1),\ F_{4(-20)} \end{cases} \tag{2.7}$$

For G=SU(p,1) with p even.

$$I(s;b) \stackrel{\text{def}}{=} C_G \pi \sum_{j=0}^{p-1} a_{2j}\, j! \left[\frac{\pi}{4} j! \sum_{\ell=0}^{j} \frac{K_{j-\ell}(s-\ell-1; b+\rho_0, \frac{\pi}{2})}{(j-\ell)!(s-1)(s-2)\ldots(s-(\ell+1))} + L_j(s; b+\rho_0, \frac{\pi}{2})\right] \tag{2.8}$$

where $\rho_0^2 = p^2$ by (2.5). For G = SO(2n+1,1)

$$I(s;b) \stackrel{def}{=} \frac{C_G\pi}{\Gamma(s)}\sum_{j=0}^{n} a_{2j}(b+\rho_0^2+r^2)^{j+\frac{1}{2}-s}\Gamma(j+\tfrac{1}{2})\Gamma(s-j-\tfrac{1}{2}) \tag{2.9}$$

where $\rho_0^2 = n^2$ by (2.5). The function I(s;b) provides for the meromorphic continuation of the function $s \to \int_R \frac{|c(r)|^{-2}}{(b+\rho_0^2+r^2)^s} dr$ for $|c(r)|^2$ in (2.1):

Theorem 2.10([36], or [41]) $s \to \int_R \frac{|c(r)|^{-2}}{(b+\rho_0^2+r^2)^s} dr$ is a well defined holomorphic function on $\mathrm{Re}\, s > \frac{d}{2}$, and coincides with the function I(s;b) on this domain.

3. The Minakshisundaram-Pleijel zeta function and Selberg zeta function related

In this section we present the basic result on the meromorphic continuation of the M-P spectral zeta function $D_\Gamma(s;\chi)$ given in (1.3), especially on the domain $\mathrm{Re}\, s < 0$ on which $D_\Gamma(s;\chi)$ is expressed directly in terms of the logarithmic derivative $\Psi_\Gamma(s;\chi)$ of Selberg's zeta function $Z_\Gamma(s;\chi)$ (and the spherical harmonic analysis of the symmetric space G / K). From this result one immediately computes the vacuum energy $E_v(\Gamma,\chi)$ for the Clifford-Klein space-times $X_\Gamma = R \times (\Gamma \setminus G / K)$ (for G given in (1.1) up to local isomorphism). For wider applications we will work in fact with spectral zeta functions

$$D_\Gamma(s;b,\chi) = \sum_j{}' \frac{n_j(\chi)}{(b+\lambda_j)^s} \tag{3.1}$$

with the parameter $b \geq 0$, where the prime means that the summation is from j = 1 to ∞ if b = 0 and from j = 0 to ∞ if b > 0; $D_\Gamma(s;\chi) = D_\Gamma(s;0,\chi)$.

Given the data (G, K, Γ, χ), the meromorphic function $\Psi_\Gamma(s;\chi)$ (which is holomorphic on the domain $\mathrm{Re}\, s > 2\rho_0$ for ρ_0 in (2.5)) is constructed by Gangolli in [18], using the trace formula. A certain constant κ is employed in [18] in setting up $\Psi_\Gamma(s;\chi)$, but as shown in [16] one may take κ = 1. One has a meromorphic function $Z_\Gamma(s;\chi)$ satisfying $Z_\Gamma'(s;\chi) / Z_\Gamma(s;\chi) = \Psi_\Gamma(s;\chi)$. $Z_\Gamma(s;\chi)$ suitably normalized is the Selberg zeta function attached to the data(G, K, Γ, χ) [4], [16], [18], [19], [21], [24], [31], [32], [34], [37], [38], [40].

For $b \geq 0$ define a domain D(b) by

$$D(b) \stackrel{def}{=} \{s \in C \mid \mathrm{Re}\, s < 1\} \text{ if } b > 0$$

$$\text{or } \{s \in C \mid \mathrm{Re}\, s < 0\} \text{ if } b = 0 \tag{3.2}$$

and set

$$I_\Gamma(s;b,\chi) \stackrel{def}{=} \int_0^\infty \psi_\Gamma(\rho_0 + t + a(b);\chi)(2a(b)t + t^2)^{-s}dt \tag{3.3}$$

for $s \in D(b)$, where $a(b) \stackrel{def}{=} +\sqrt{b+\rho_0^2}$.

Proposition 3.4([42]) The improper integral in (3.3) converges absolutely on the domain D(b) in (3.2) and defines a holomorphic function of s on this domain.

Here's the main theorem:

Theorem 3.5([42]) For $b \geq 0$ let $D_\Gamma(s;b,\chi)$ be the spectral zeta function of Minakshisundaram-Pleijel type given in (3.1). Then $D_\Gamma(s;b,\chi)$ converges absolutely for Re $s > \frac{d}{2}$ (see (2.4)), is holomorphic in s on this domain and admits a meromorphic continuation to the full complex plane:

$$D_\Gamma(s;b,\chi) = c\chi(1) \text{ vol } (\Gamma \setminus G)\, I(s;b) + E_\Gamma(s;b,\chi) \text{ for Re} s > \frac{d}{2} \tag{3.6}$$

for a suitable entire function $E_\Gamma(s;b,\chi)$ where $I(s;b)$ is given by (2.7), (2.8), or (2.9) for the various cases of G in (1.1); c is a suitable constant. In particular on the domain D(b) in (3.2)

$$E_\Gamma(s;b,\chi) = \frac{(\sin \pi s)}{\pi} I_\Gamma(s;b,\chi) \tag{3.7}$$

where $I_\Gamma(s;b,\chi)$ is given in (3.3); see Proposition 3.4.

$c = \frac{1}{4\pi}$ for the normalization of measures in [36], [41]. In (3.6), $\text{vol}(\Gamma \setminus G) \stackrel{def}{=} \int_{\Gamma\setminus G} 1\, dm_\Gamma$ as usual, where dm_Γ is G - invariant measure on $\Gamma \setminus G$ induced by a Haar measure on G. From (2.7), (2.8), (2.9) we see that apart from the case $G = SO_1(n,1)$ with n odd, $D_\Gamma(s;b,\chi)$ is holomorphic except for possibly a simple pole at s = 1, 2, 3,..., or $\frac{d}{2}$. In particular we have holomorphicity at $s = -\frac{1}{2}$, which means that the parametrized vacuum energy

$$E_v(\Gamma, \chi; b) \stackrel{def}{=} \frac{1}{2} D_\Gamma(-\tfrac{1}{2}; b,\chi) \tag{3.8}$$

(cf. remarks following (1.3)) for free massless scalar fields over our ultrastatic space-times $X_\Gamma = R \times (\Gamma \setminus G / K)$ (with $G \neq SO_1(n,1)$ for n odd) is well-defined and finite, and is given by

Corollary 3.9

$E_v(\Gamma,\chi;b) = \frac{c}{2}\chi(1)\ \mathrm{vol}(\Gamma\setminus G)\ I(-\tfrac{1}{2};b) - \frac{1}{2\pi}\int_0^\infty \Psi_\Gamma(\rho_0 + t + a(b);\chi)(2a(b)t + t^2)^{\frac{1}{2}}dt$ for $a(b) = +\sqrt{b+\rho_0^2}$, where $I(-\frac{1}{2};b)$ is given by (2.7) or (2.8). In particular for the trivial representation $\chi = 1$ of Γ the topological component of $E_v(\Gamma,\chi;b)$ (i.e. the term $-\frac{1}{2\pi}\int_0^\infty \Psi_\Gamma(\rho_0 + t + a(b);\chi)(2a(b)t + t^2)^{\frac{1}{2}}dt$) is always negative.

Corollary 3.9 extends to the general rank 1 symmetric space $X = G / K$ work of Bytsenko, Goncharov, and Zerbini [9] who concentrated on the case $X = X(n) \stackrel{def}{=} SO_1(n,1)/SO(n)$; see [7], [8], [14] for the cases $X(2)$, $X(3)$. Since $a(0) = \rho_0$ we see that

$$E_v(\Gamma,\chi) = \frac{c}{2}\chi(1)\ \mathrm{vol}(\Gamma\setminus G)\ I(-\tfrac{1}{2};0) - \frac{1}{2\pi}\int_0^\infty \Psi_\Gamma(t + 2\rho_0;\chi)(2\rho_0 t + t^2)^{\frac{1}{2}}dt \tag{3.10}$$

In the sole case $G = SO_1(n,1)$ with n odd (which is exactly the case when dim X is odd), $D_\Gamma(s;b,\chi)$ has a simple pole at $s = -\frac{1}{2}$ (by (2.9)) and hence definition (3.8) needs to be replaced by a more general definition, as indicated earlier. Physicists are also interested in the computation of $E_v(\Gamma,\chi)$ in cases where Γ, more generally, might have elliptic elements. Here again the trace formula plays the key role.

4. Remarks on the case of non-compact $\Gamma \setminus G$

So far we have imposed a compactness assumption on the quotient $\Gamma \setminus G$. Assume now that $\Gamma \setminus G$ is non-compact but $\mathrm{vol}(\Gamma \setminus G) < \infty$; cf. remark following (3.7). In this case the notion of a fully discrete spectrum $\{\lambda_j, n_j\}_{j=0}^\infty$ of $-\Delta_\Gamma$ in the sense considered earlier no longer prevails. We can however reinterpret this spectrum in representation - theoretic terms, which have a meaning for $\Gamma \setminus G$ non-compact. Thus we can construct a M-P type zeta function $D_\Gamma(s;b)$ (where we specialize χ to be the trivial representation of Γ in this section) and consider its meromorphic continuation - as an interesting mathematical problem in its own right, and as a response to the issue raised in [9] regarding the regularization of vacuum energy E_v for $\Gamma \setminus G$ non-compact. We find that, in contrast to the compact case, $D_\Gamma(s;b)$ has double poles at the points $s_k = \frac{1}{2} - k$, $k = 0, 1, 2,\ldots$ and thus at $s = -\frac{1}{2}$ in particular. The regularization based on (1.2) (evaluating $D_\Gamma(s;b)$ at $s= -\frac{1}{2}$, say for dim X even) therefore cannot be employed. The existence of the double poles s_k (already noticed by Kurokawa in the lowest dimensional case $G = SO_1(2,1) \approx SL(2,R)$, $\Gamma=SL(2,Z)$ [24]) is due to a parabolic contribution to the trace formula.

For $\Gamma \setminus G$ non-compact the Hilbert space $L^2(\Gamma \setminus G)$ as a unitary representation of G (where G acts by right translation on functions) has a G-orthogonal decomposition

$$L^2(\Gamma \setminus G) = L^2_{Eis}(\Gamma \setminus G) \oplus L^2_{dis}(\Gamma \setminus G) \tag{4.1}$$

by Selberg-Langlands theory [25], [29], [31], [33] where $L^2_{Eis}(\Gamma \setminus G)$ is the closed subspace of $L^2(\Gamma \setminus G)$ generated by wave packets of Eisenstein series, and $L^2_{dis}(\Gamma \setminus G)$ is the orthogonal direct sum of closed subspaces L^2_{cusp}, L^2_{res} of $L^2(\Gamma \setminus G)$ generated by cusp forms and square-integrable residues of Eisenstein series. Moreover

$$L^2_{dis}(\Gamma \setminus G) = \sum_{\pi \in \hat{G}} \oplus\, m_\pi(\Gamma) H_\pi \tag{4.2}$$

is a discrete orthogonal direct sum over the unitary dual $\hat{G}$ of G, where each irreducible unitary representation π of G (on the Hilbert space H_π) occurs with finite (possibly zero) multiplicity $m_\pi(\Gamma)$. Here Γ is subject to mild generic conditions. For example, we assume for simplicity that Γ satisfies the following two conditions (as we shall need to apply results from [33]) : (i) torsion in Γ is confined to its center $z(\Gamma)$ - i.e. only central elements of Γ have finite order, (ii) $\Gamma \cap P_j = z(\Gamma)(\Gamma \cap N_j)$ for $1 \leq j \leq c$ where $\{P_1, \ldots, P_c\}$ is a complete set of representatives of the Γ-cuspidal minimal parabolic subgroups of G. Here $P_j = M_j A_j N_j$ is a Langlands decomposition of P_j; we take $P_1 = MAN$ for M = the centralizer A in K, given the Iwasawa decomposition KAN of G as in section 2. Γ-cuspidality means that $(\Gamma \cap N_j) \backslash N_j$ is compact.

Fix a complete list $\{\pi_j\}_{j=0}^{\infty}$, up to unitary equivalence, of the irreducible unitary class 1 representation of G (i.e. $\pi_j|_K$ contains the trivial representation of K) which occur in $L^2_{dis}(\Gamma \setminus G)$:

$$m_j \overset{def}{=} m_{\pi_j}(\Gamma) > 0 \text{ in } (4.2) \tag{4.3}$$

Being a class 1 representation, π_j corresponds to a Harish-Chandra spherical function $\Phi = \Phi_{v_j}$ on G for a suitable v_j in the complexified dual a_0^{*C} of the Lie algebra a_0 of A. Since $\dim a_0 \overset{def}{=} \text{rank of } X = 1$ we consider v_j as an element in C. The v_j are determined up to the action of the Weyl group of (G,A), and we may label them so that the

$$\lambda_j \overset{def}{=} v_j + \rho_0^2 \tag{4.4}$$

for ρ_0 in (2.5), satisfy $0 = \lambda_0 < \lambda_1 < \ldots$; $\lim_{\lambda_j \to \infty} \lambda_j = \infty$. In fact for $\Gamma \setminus G$ compact one knows that the λ_j in (4.4) are the eigenvalues λ_j of $-\Delta_\Gamma$ in section 1 and the multiplicity m_j in (4.3) coincides with the multiplicity n_j; here $L^2_{dis}(\Gamma \setminus G)$ (see(4.2)) $\equiv L^2(\Gamma \setminus G)$. It follows that (especially for $\Gamma \setminus G$ non-compact) the role of the spectrum $\{\lambda_j, n_j\}$ of $-\Delta_\Gamma$ in section 1 is subsumed by the data $\{\lambda_j, m_j\}$ of (4.3), (4.4), and that we can form the following M-P type zeta function

$$D_\Gamma(s; b) = \sum_j{}' \frac{m_j}{(b+\lambda_j)^s} \tag{4.5}$$

for $\mathrm{Re}\, s >> 0$, where the summation is from $j = 1$ to ∞ if $b = 0$ and from $j = 0$ to ∞ if $b>0$; cf. (3.1).

For $b \geq 0$ we have the ϑ-function $\vartheta_\Gamma(t;b)$ with parameter b given by

$$\vartheta_\Gamma(t;b) \overset{\mathrm{def}}{=} \sum_{j=0}^{\infty} m_j e^{-(\lambda_j+b)t} = e^{-bt}\vartheta_\Gamma(t;\,0) \tag{4.6}$$

for $t > 0$ and the data (λ_j, m_j) of (4.3), (4.4). Let $S = [S_{ij}]$ be the scattering matrix associated to the theory of spherical Eisenstein series of (G, Γ); see [19], [40] for details where S_{ij} there is denoted by M_{ij}. Recalling that c denotes the number of Γ-parabolic cusps (see condition (ii) on Γ), S is a c^2-matrix of meromorphic functions S_{ij} on $a_0^{*\,C} \approx C$; again use rank $X = 1$ for the latter isomorphism. A key function is the meromorphic function

$$D(z) \overset{\mathrm{def}}{=} \det S(z) \tag{4.7}$$

for $z \in C$. In terms of D and Harish-Chandra's density $|c(\,r)|^{-2}$ in (2.1) the heat equation version of the trace formula (due to Warner [19], [33]) takes the form (see(4.6))

$$\vartheta_\Gamma(t;\,0) = c_1|z(\Gamma)|\ \mathrm{vol}(\Gamma\setminus G)\int_R e^{-(r^2+\rho_0^2)t}|c(r)|^{-2}\,dr \ + \text{the hyperbolic contribution}$$

$$+\, c_2\int_R e^{-(r^2+\rho_0^2)t} D'(ir)/D(ir)dr - \frac{c}{2\pi}\int_R e^{-(r^2+\rho_0^2)t}\Gamma'(1+ir)/\Gamma(1+ir)dr$$
$$+\, c_3\int_R e^{-(r^2+\rho_0^2)t}dr + c_4\int_R e^{-(r^2+\rho_0^2)t}J(r)dr + \tfrac{1}{4}(c - \mathrm{trace}\ S(0))\, e^{-\rho_0^2 t} \tag{4.8}$$

for suitable constants c_1, c_2, c_3, c_4 where J is an entire function and the 6th term involving it occurs in cases (ii), (iii), (iv) in (1.1) and does not appear in the case $G = SO(n,1)$. Thus $J(\,r)$ does not appear in Selberg's paper [31]. In the former cases (ii), (iii), (iv), J is in fact an even polynomial except when $G = SU(n,1)$ with n even (the cotangent case in (2.1)). For $\Gamma \setminus G$ compact only the first two terms of (4.8) (the central and hyperbolic contributions) appear. Details on this appear in [31], [33], [35], [40].

The basic result of this section is the following theorem which can be regarded as a generalization of Theorem 3.5, where we assume $b > 0$ for simplicity and where we exclude the cotangent case $G = SU(n,1)$, n even:

Theorem 4.9 For $t > 0$, let

$$E_\Gamma(t;b) \underset{=}{\text{def}} e^{-bt}(3^{rd} \text{ terms in } (4.8)) = c_2 \int_R e^{-(r^2+\rho_0^2+b)t} D'(ir)/D(ir)\,dr \quad (4.10)$$

for D in (4.7). Then the Mellin transform

$$M_\Gamma(s;b) \underset{=}{\text{def}} \int_0^\infty [\vartheta_\Gamma(t;b) - E_\Gamma(t;b)]t^{s-1}dt \quad (4.11)$$

exists for $\text{Re}\,s > \frac{d}{2}$ ($d = \dim X$; see (2.4)), is holomorphic in s on this domain, and admits a reasonably explicit meromorphic continuation to the full complex plane C, with computable poles and residues. In particular $M_\Gamma(s;b)/\Gamma(s)$ is holomorphic at $s = 0$ and has double poles at $s_k \underset{=}{\text{def}} \frac{1}{2}-k$, $k=0,1,2,3,\ldots$, in addition to possible simple Minakshisundaram-Pleijel poles at $s=1, 2,\ldots \frac{d}{2}$.

As in Theorem 3.5 the harmonic analysis of X figures prominently in the meromorphic continuation of $M_\Gamma(s;b)$. The existence of the double poles s_k in Theorem 4.9 (whose implication for physics has already been remarked on) is due to the contribution of the logarithmic derivative $\Psi(z) = \Gamma'(z)/\Gamma(z)$ of the gamma function $\Gamma(z)$ to the trace formula - the 4th term in (4.8). Namely, we take the Mellin transform of this 4th term and interchange the order of integration to obtain $-\frac{c}{2\pi}\Gamma(s)\int_R \frac{\Psi(1+ir)}{(r^2+\rho_0^2)^s}dr$, where the latter integral as a function of s can be related to the function Q(s) in [24], which is shown in [24] to be holomorphic except for the points s_k which are double poles. The proof of Theorem 4.9 proceeds by analyzing similarly the Mellin transform of the other terms in (4.8), apart from the 3rd term.

To understand the meaning of Theorem 4.9 a bit better note first that in general the scattering matrix S is unknown in explicit terms for most (G, Γ), even for $G = SL(2, R)$. In cases where S (or D) is known one has an estimate

$$|D'(ir)/D(ir)| \leq \text{constant}\,|r|^k \quad (4.12)$$

say with $1 \leq k+1 \leq d$. Such an estimate generally involves estimates on Riemann's zeta function $\zeta(s)$ and various Dirichlet L-functions. For example, for $(G,\Gamma) = (SL(2, R), SL(2,Z))$, $D(s)=\sqrt{\pi}\,[\zeta(2s)/\zeta(2s+1)]\Gamma(s)/\Gamma(s+\frac{1}{2})$ [40]. Given (4.12) the Mellin transform

$$\hat{E}_\Gamma(s;b) \underset{=}{\text{def}} \int_0^\infty E_\Gamma(t;b)t^{s-1}dt \text{ of } E_\Gamma(t;b) \text{ in } (4.10)$$

exists for Res > $\frac{k+1}{2}$ (and thus in particular for Res > $\frac{d}{2}$). The problem remains to meromorphically continue $\hat{E}_\Gamma(s;b)$ in s given knowledge of D for a specific pair (G, Γ). Then by Theorem 4.9

$$\hat{\vartheta}_\Gamma(s;b) \underset{=}{\text{def}} \int_0^\infty \vartheta_\Gamma(t;b)t^{s-1}dt \tag{4.13}$$

is holomorphic for Res > $\frac{d}{2}$ and admits a meromorphic continuation to C. But

$$\hat{\vartheta}_\Gamma(s;b) = \sum_{j=0}^\infty m_j \int_0^\infty e^{-(\lambda_j+b)}t^{s-1}dt = \sum_{j=0}^\infty \frac{m_j}{(\lambda_j+b)^s}\Gamma(s) = D_\Gamma(s;b)\Gamma(s) \quad \text{for Res} > \frac{d}{2} \tag{4.14}$$

by (4.5), (4.6)! Thus in principle, Theorem 4.9 provides indeed for the meromorphic continuation of $D_\Gamma(s;b)$, and for the computation of its poles and residues. Other applications of Theorem 4.9 or Theorem 3.5 include a factorization on Selberg's zeta function $Z_\Gamma(s;\chi)$ and a new simple proof of Gelfand's conjecture (Weyl's law) on the asymptotic behavior of the class 1 spectrum $\{\lambda_j, m_j\}_{j=0}^\infty$ [39].

Summary

Using Selberg's trace formula we have expressed in Theorem 3.5 a direct relationship between the Minakshisundaram-Pleijel spectral zeta function and the Selberg zeta function of a compact rank 1 space-form of a non-compact symmetric space *X*. From this result we obtained, in particular, an explicit formula for the Casimir energy of a free massless scalar field over an ultrastatic space-time modeled on *X*, generalizing work of Bytsenko, Goncharov, Zerbini, and others. We have also shown that for a non-compact space form of *X* the spectral zeta function has a double pole at s = $-\frac{1}{2}$, and thus (in this case) physicists will need to consider a modified regularization procedure in order to obtain a finite Casimir energy.

REFERENCES

[1] R. Banach and J. Dowker, Quantum field theory on Clifford-Klein space-times. The effective Lagrangian and vacuum stress-energy tensor, J.Phys. A 11, 2255- 2284 (1978).

[2] R. Banach and J. Dowker, The vacuum stress tensor for automorphic fields on some flat space-times, J. Phys. A 12, 2545- 2562 (1979).

[3] N. Birrell and P. Davies, Quantum fields on curved spaces, Cambridge University Press, Cambridge, England (1982).

[4] U. Bunke and M. Olbrich, Selberg zeta and theta functions - a differential operator approach, Math. Research Vol. 83, Akademic Verlag, Berlin (1995).

[5] A. Bytsenko, G. Cognola, L. Vanzo, and S. Zerbini, Quantum fields and extended objects in space-times with constant curvature spatial section, Phys. Reports, Vol. 226, 1-126 (1996).

[6] A. Bytsenko, E. Elizalde, S. Odintsov, A. Romeo, and S. Zerbini, Zeta regularization techniques with applications, World Scientific, Singapore (1994).

[7] A. Bytsenko and Y. Goncharov, Topological Casimir effect for a class of hyperbolic three-dimensional Clifford-Klein space-times, Classical Quantum Gravity 8, 2269-2275 (1991). Also Modern Phys. Letters A6, 669 (1991).

[8] A. Bytsenko and Y. Goncharov, Topological Casimir effect for a class of hyperbolic four- dimensional Clifford-Klein space-times, Letter to the Editor, Classical Quantum Gravity 8, L211-L214 (1991).

[9] A. Bytsenko and S. Zerbini, The Casimir effect for a class of Hyperbolic (D+1)-dimensional space-times, a Classical Quantum Gravity 9, 1365-1370 (1992).

[10] R. Camporesi, Harmonic analysis and propagators on homogeneous spaces, Phys. Reports, Vol. 196, 1-134 (1990).

[11] H. Casimir, Proc. Kon. Ned. Akad., Wetenschap 1351, 793 (1948). Also Physica 19, 846 (1953).

[12] Harish-Chandra, Spherical functions on semisimple Lie groups I, II, Am. J. Math. 80, 241-310, 553-613 (1958).

[13] Harish-Chandra, Discrete series for semisimple Lie groups II, Acta Math. 116, 1-111 (1966).

[14] G. Cognola, L. Vanzo, and S. Zerbini, Regularization dependence of vacuum energy in arbitrarily shaped cavities, J. Math, Phys. 33, 222-228 (1992).

[15] E. Elizalde, Ten physical applications of spectral zeta functions, Lecture Notes in Physics m 35, Springer-Verlag (1995).

[16] D. Fried, The zeta functions of Ruelle and Selberg I, Ann. Sci. E'cole Norm. Sup. 10, 133-152 (1977).

[17] S. Fulling, Non-uniqueness of canonical field quantization in Riemannian space-time, Phys. Review D, Vol. 7, 2850-2862 (1973).

[18] R. Gangolli, Zeta functions of Selberg's type for compact space forms of symmetric spaces of rank one, Illinois J. Math. 21, I-42 (1977).

[19] R. Gangolli and G. Warner, Zeta functions of Selberg's type for some non-compact quotients of symmetric spaces of rank one, Nagoya Math. J. 78, 1-44 (1980).

[20] S. Hawking, Zeta function regularization of path integrals in curved space-time, Comm. Math. Phys. 55, 133-148 (1977).

[21] D. Hejhal, The Selberg trace formula for SL(2,R), I, II, Lecture Notes in Math. Springer-Verlag, 548 (1976) and 1001 (1983).

[22] S. Helgason, Differential geometry and symmetric spaces, Pure and Applied Math. Ser. 12, Academic Press (1962).

[23] C. Isham, Twisted quantum fields in a curved space-time, Proc. Royal Soc. London A 362, 383-404 (1978).

[24] N. Kurokawa, Parabolic components of zeta functions, Proc. Japan Acad. 64A, 21-24 (1988).

[25] R. Langlands, On the functional equation satisfied by Eisenstein series, Lecture Notes in Math. 544, Springer-Verlag (1976).

[26] R. Miatello, The Minakshisundaram-Pleijel coefficients for the vector-valued heat kernel on compact locally symmetric spaces of negative curvature, Trans. Am. Math. Soc. 260, 1-33 (1980).

[27] R. Miatello, On the Plancherel measure for linear Lie groups of rank one, Manuscripta Math. 29, 249-276 (1979).

[28] S. Minakshisundaram and A. Pleijel, Some properties of the eigen functions of the Laplace operator on Riemannian manifolds, Canadian J. Math. 1, 242-256 (1949).

[29] S. Osborne and G. Warner, The theory of Eisenstein systems, Pure and Applied Math. 99, Academic Press (1981).

[30] B. Randol, On the analytic continuation of the Minakshisundaram-Pleijel zeta function for compact Riemann surfaces, Trans. Am. Math. Soc. 201, 241-246 (1975).

[31] A. Selberg, Harmonic analysis on discontinuous groups in weakly symmetric Riemannian spaces with applications to Dirichlet series, J. India math; Soc. 20, 47-48 (1956).

[32] D. Scott, Selberg type zeta functions for the group of complex two by two matrices of determinant one, Math. Ann. 253, 177-194 (1980).

[33] G. Warner, Selberg's trace formula for non-uniform lattices-the R-rank one case, Advances in Math. Studies 6, 1-142 (1979).

[34] M. Wakayama, Zeta functions of Selberg's type associated with homogeneous vector bundles, Hiroshima Math. J. 15, 235-295 (1985).

[35] N. Wallach, On the Selberg trace formula in the case of compact quotient, Bull. Am. Math. Soc. 82, 171-195 (1976).

[36] F. Williams, Formula for the class 1 spectrum and the meromorphic continuation of Minakshisundaram-Pleijel series preprint (1989).

[37] F. Williams, A factorization of the Selberg zeta function attached to a rank 1 space form, Manuscripta Math. 77, 17-39 (1992).

[38] F. Williams, Some zeta functions attached to $\Gamma \backslash G / K$, in New Developments in Lie Theory and Their Applications, edited by J. Tirao and N. Wallach, Birkhäuser Progress in Math. Ser. 105, 163-177 (1992).

[39] F. Williams, Spectral zeta series of rank 1 space forms, in Representation Theory and Harmonic Analysis, a Conference in Honor of R. Kunze, Contemporary Math. 191, 245-254 (1995).

[40] F. Williams, Lectures on the spectrum of $L^2(\Gamma \backslash G)$, Pitman Rearch Notes in Math. 242, Longman House Pub. (1990).

[41] F. Williams, Meromorphic continuation of Minakshisundaram-Pleijel series for semisimple Lie groups, 1995 rewritten version of a 1989 preprint.

[42] F. Williams, Topological Casimir energy for general class of Clifford-Klein space-times, to appear in J. Math. Physics, Feb. or March issue, 1997.

Department of Mathematics and Statistics
University of Massachusetts
Amherst, MA 01003-4515
williams@math.umass.edu

DIMACS Series in Discrete Mathematics
and Theoretical Computer Science
Volume **34**, 1997

Some Dynamics on the Irrationals

Scott W. Williams

ABSTRACT. After observing that a discrete dynamical system on a separable completely metrizable space is the homomorphic image of a dynamical system on the space of irrationals, we give a unified construction of many examples of dynamical systems on the space of the irrationals.

0. Introduction

We present here work which is, in part, expository with proofs, exercises (2.4, 2.6, 4.3, and 7.6), and, in part, contains new results (3.1, 5.4, and 7.7) so we ought to begin with some background: Dynamics travels a line of history from as far back as Newton, as a notion for his laws of motion. Prior to the twentieth century, a dynamical system meant a motion whose parameters are functions of time and satisfy a system of differential equations. Eighteenth and Nineteenth century analysts used various analytical manipulations (including infinte series) to cause the differential equations to reveal information. However, 100 years ago Poincaré, using proofs foretelling modern topology, shifted our attention from particular solutions to the relationships between all possible solutions and, in some cases, he used his methods to prove the existence of periodic solutions. In 1927, G.D. Birkhoff's work signifcantly justified Poincaré's global approach by proving that every dynamical system on a compact space has a solution stable in the sense of Poisson. In our terminolgy we would state Birkoff's result as "each system on a compact space has a recurrent point." In the area of multiple systems, the first result, "if x is recurrent in a system $[X;f]$, then $\forall n \in \mathbf{N}$, x is recurrent in $[X;f^n]$," and came in with a little heralded paper [8] of P. Erdös and A. H. Stone. H. Furstenburg, B. Weiss, and others built, in the late 1960's, a theory for multiple systems (see section 7), paralleling Birkhoff's for simple systems.

$\mathbf{P}$ denotes the subspace of irrationals in the real line, while Π denotes a certain space (yet to be defined) homeomorphic to $\mathbf{P}$. The sections of this paper are:

0. Introduction. 1. A homeomorph of $\mathbf{P}$. 2. A new system on $\mathbf{P}$. 3. More homeomorphs of $\mathbf{P}$. 4. Constructing points in Π. 5. Recurrent points in $[\Pi;\sigma]$. 6. Systems with all points recurrent. 7. Multiple Recurrence.

1991 *Mathematics Subject Classification.* 54H20, 54D40.

In general, the objects considered here are pairs $[X;f]$, called a discrete dynamical system, where X is a metric space and $f : X \to X$ is a continuous function (It is common, but not necessary, to require f to be a homeomorphism, and I assure you all the material here have analogous results with much more complex proofs, in the homeomorphism case, see [16].). Our particular attention will focus upon the case when X is one of the topological spaces $\mathbf{P}$ or Π.

The primary results we exhibit are:

3.1. Suppose $[X,f]$ is a system on a separable complete metrizable space (X,d). Then there is a system $[\mathbf{P},g]$ and a homomorphism $h: [\mathbf{P},g] \to [X,f]$.
5.2. Suppose $x \in \Pi$. x is recurrent in $[\Pi;\sigma]$ iff there is a street S with $x = \alpha S$.
5.3. If $\{\lambda B_z : z \in \mathbf{N}\}$ is finite, then x is almost-periodic in $[\Pi;\sigma]$.
5.4. There is an almost periodic-point x such that if $(F_z : z \in \mathbf{N})$ is a street with $x = \alpha(F_z : n \in \mathbf{N})$. Then $\{\lambda F_z : z \in I\}$ is infinite.
6.3. There is a homeomorphism $f: \mathbf{P} \to \mathbf{P}$ such that $[\mathbf{P};f]$ is a minimal system, and no point of P is almost-periodic in $[\mathbf{P};f]$.
6.4. There is a continuous function $f: \mathbf{P} \to \mathbf{P}$ such that each point in $\mathbf{P}$ is recurrent in $[\mathbf{P};f]$, but $[\mathbf{P};f]$ has no minimal sets.
7.7. There is a continuous function $f: \mathbf{P} \to \mathbf{P}$ such that $[\mathbf{P},f]$ is minimal, but no point of $\mathbf{P}$ is multiply recurrent in $[\mathbf{P};\{f,f^2\}]$.
7.9. There are commuting continuous functions $f,g: \mathbf{P} \to \mathbf{P}$ such that $[\mathbf{P},\{f,g\}]$ is minimal, $\forall p \in \mathbf{P}$, $\mathbf{OC}_f(p) \cap \mathbf{OC}_g(p) = \{p\}$.

All the definitions are standard (see [1], [6], [11], [12], and [17]). Fix a system $[X;f]$, let $f^1 = f$, and for each $n \in \mathbf{N}$ (= the set of positive integers) let $f^{n+1} = ff^n$, the composition of f with f^n. The *orbit* of a point $x \in X$ is the set $\{f^n(x) : n \in \mathbf{N}\}$, denoted by $\mathbf{Orb}_f(x)$ (or $\mathbf{Orb}(x)$ when no confusion results). The *orbit closure* of a point $x \in X$ is the set $cl(\mathbf{Orb}(x))$, denoted by $\mathbf{OC}_f(x)$ (or $\mathbf{OC}(x)$ when no confusion results). A set (or the system) is said to be *transitive* if it equals some $\mathbf{OC}(x)$.

Suppose $x \in X$. In $[X;f]$,

1). x is said to be *fixed* provided $f(x) = x$.
2). x is said to be *periodic* provided $\exists m \in \mathbf{N}$ such that $f^m(x) = x$. (of period $m \in \mathbf{N}$, if m is the first such integer).
3). A point $x \in X$ is said to be *almost-periodic* (also known as *uniformly recurrent* [9]) in $[X;f]$ provided for each neighborhood U of x, $\{n \in \mathbf{N} : f^n(x) \in U\} \neq \emptyset$ is relatively dense in $\mathbf{N}$; i.e., provided $\exists k = k(U) \in \mathbf{N}$ such that $\forall m \in \mathbf{N}$, $\{n \in \mathbf{N} : km \leq n < (k+1)m,\ f^n(x) \in U\} \neq \emptyset$.
4). A point $x \in X$ is said to be *recurrent* in $[X;f]$ provided $x \in \mathbf{OC}(x)$ (or equivalently, when X is a metric space, there is an increasing sequence $\langle k_n \rangle$ in $\mathbf{N}$ such that $\lim_{n\to\infty} f^{k_n}(x) = x$).
5). A set $M \subseteq X$ is said to be *minimal* in $[X;f]$ provided it is a minimal element in the partially ordered, by $\subseteq$, set of all non-empty closed sets $A \subseteq X$ such that $f(A) \subseteq A$, or equivalently, if $\forall x \in M$, $\mathbf{OC}(x) = M$. $[X;f]$ is called a *minimal system* when X is minimal.

There are more interesting points we could study here (see exercise 2.6 and [1], [2], [6], [9], [10], [12], [15], [17], and [18]).

In dynamics, the basic property preserving properties between two systems are called homomorphisms. Specifically, suppose [X,f] and [Y,g] are systems. h is a *homomorphism* from [Y,g] to [X,f] (write $h : [Y,g] \to [X,f]$) if $h : Y \to X$ is a continuous surjection satisfying $gh = hf$. It is very easy to show a homomorphism takes fixed (periodic, almost-periodic, recurrent) points to fixed (respectively, periodic, almost-periodic, recurrent) points, and that the composition of two homomorphisms is a homomorphism.

In the final section of this paper, we will consider *multiple systems* $[X;\mathfrak{F}]$, where $\mathfrak{F}$ is a commuting family ($\forall f,g \in \mathfrak{F}$, $fg = gf$) of continuous functions $f : X \to X$. Given a multiple system $[X;\mathfrak{F}]$, let $<\mathfrak{F}>$ be the semi-group generated by $\mathfrak{F}$; i.e., $\{f_1 f_2 \cdots f_k : <f_1, f_2, \ldots, f_k> \text{ is a finite sequence in } \mathfrak{F}\}$.

For $x \in X$, we set $\mathbf{OC}_{\mathfrak{F}}(x) = \text{cl}\{fx : f \in <\mathfrak{F}>\}$. A *minimal* set in a multiple system $[X;\mathfrak{F}]$ is a set $M \subseteq X$ minimal in the partially ordered, by $\subseteq$, set of all non-empty closed sets $A \subseteq X$ such that $\forall f \in \mathfrak{F}$, $fA \subseteq A$.

$x \in X$ is said to be *jointly recurrent (almost-periodic)* in the multiple system $[X;\mathfrak{F}]$ provided $\forall f \in \mathfrak{F}$, x is recurrent (almost-periodic) in [X;f]. In the case $\mathfrak{F}$ is finite, x is said to be *multiply recurrent* [9] in $[X;\mathfrak{F}]$ provided that for each neighborhood U of x $\exists n \in \mathbf{N}$ such that $\forall f \in \mathfrak{F}$, $f^n x \in U$ [9] (this is usually defined when $\mathfrak{F}$ is finite; the infinite case is considered in [2] where it is called *uniform multiple recurrence*).

C, **N**, **Q**, **R**, **Z**, and ω denote, respectively, the Cantor middle-thirds set, the positive integers, the space of rationals, the reals, the integers, and the non-negative integers. card(X) denotes the cardinality of a set X.

Given a space X and a set I, we use Logic's notation ${}^I X$ to denote all functions from I to X. $\Pi^I X$ denotes the Tychonov product of I many copies of X. Π will be used to denote the so called Baire space of all sequences $<x_n>$ of positive integers with the metric $d(x,y) = 2^{-(n-1)}$, if n is the least integer such that $x_n \neq y_n$.

cl and int denote, respectively, the closure and interior operators in a space X. **C**, **N**, **P**, **Q**, **R**, **Z**, and ω all possess natural linear orders and we use [a,b] and (a,b), respectively, to denote closed and open intervals.

A *zero-dimensional* space is a space with a base of *clopen* (≡ simultaneously open and closed) sets. A space X is *separable* provided it has a countable dense set; i.e., there is a countable subset of X whose closure is X. A space is *completely metrizable* provided it is homeomorphic to a complete metric space. A space is *nowhere-locally compact* provided no non-empty open set has compact closure.

The *domain* and *range* of a function f are denoted, repectively, by dom and rng. **HENCEFORTH**, we use fx instead of f(x) for the image of a point unless some confusion would result.

\\ denotes the beginning of a proof, while // denotes the end.

1. A homeomorph of P

P and **C** are often considered bizarre, however, they occupy a place fundamental to dynamics, both in theory and example. Our primary goal in this paper is to present examples on **P**. The chief method is by the way of constructing "simple" systems on objects topologically the same (i.e., homeomorphic to) as **P**.

1.1. The usual topology on P. A basic nhbd of a point x has, in **P**, form $(x-\varepsilon,x+\varepsilon)$ (in **P** of course), where $\varepsilon > 0$. In **R** there is a rational between any pair of irrationals. Hence, we have alternate basic nhbds of form $(a,b)\cap\mathbf{P}$, where a and b are rationals. When p is a rational $(p,\infty)\cap\mathbf{P} = [p,\infty)\cap\mathbf{P}$. Thus, $(p,\infty)\cap\mathbf{P}$ is both open and closed in **P**. Similarly, $(-\infty,p)\cap\mathbf{P}$ is both open and closed in **P**. As each point in **P** has a nhbd base of clopen sets in **P**, **P** is a zero-dimensional space.

1.2. The usual topology on Π. **N** has the (discrete) metric

$$d(n,m) = \begin{cases} 0 & \text{if } n = m \\ 1 & \text{if } n \neq m \end{cases} .$$

Thus, the topology of $\Pi = \Pi^{\mathbf{N}}\mathbf{N}$ is also given by the complete metric $d(x,y) = 0$ if $x = y$, otherwise $d(x,y) = 2^{-k}$, where k is the first integer n such that $x(n) \neq y(n)$. Hence, Π and Π are homeomorphic, via the map which sends each sequence $<x_n>$ to the function x defined by $x(n) = x_n$. Occasionally we use this homeomorphism to identify facts about Π with corresponding facts about Π.

Suppose $F \subseteq \mathbf{N}$ is finite and $B : F \to \mathbf{N}$ is a function. Then

$$\{y \in \Pi : \forall n \in F,\ y(n) = B(n)\}$$

is an open and closed set in Π; further, the set of all such sets forms a base for the topology of Π. Therefore, a basic nhbd of $x \in \Pi$ has the form

$$\{y \in \Pi : \forall n \geq m,\ y|[1,m] = x|[1,m]\},$$

where m varies in **N**.

Usually, topology is applied to subjects such as Algebra, Analysis, or Number Theory. The next result is nearly a hundred years old and reverses this process. It was one of the most beautiful results I saw in my graduate topology course in 1965. Some of its most notable corollaries are that **P** is homeomorphic to $\mathbf{P} \times \mathbf{P}$ (the double irrationals in the plane), and that **P** has a group operation which makes it a topological group.

1.3. Theorem [7]. ***P** and Π are homeomorphic.*

\\ (Sketch of Proof) We use continued fractions: Consider sequences of the form.

$$\frac{1}{x(1)},\ \frac{1}{x(1)+\frac{1}{x(2)}},\ \frac{1}{x(1)+\frac{1}{x(2)+\frac{1}{x(3)}}},\ \ldots,$$

where $x \in {}^{\mathbf{N}}\mathbf{N}$. They define a unique real using the limit of the fractions. The function

$$f(x) = \frac{1}{x(1)+\frac{1}{x(2)+\frac{1}{x(3)+\ddots}}}$$

defines a homeomorphism between Π and $(0,1)\cap\mathbf{P}$. It is easy to see that $\mathbf{P}$ and $(0,1)\cap\mathbf{P}$ are homeomorphic. //

$F \subseteq \mathbf{N}$ is finite iff it is a compact subset of $\mathbf{N}$. Thus, Tychonov Product Theorem shows that when F is finite, $\Pi^{\mathbf{N}}F$ is a compact subset of $\Pi^{\mathbf{N}}\mathbf{N}$. When F is finite with at least two points, it can shown that $\Pi^{\mathbf{N}}F$ is homeomorphic to $\mathbf{C}$.

2. A system on P.

Applying 1.3, we may define a function from $\mathbf{P}$ to itself by defining it on Π! $\sigma : \Pi \to \Pi$ is defined by $(\sigma x)(n) = x(n+1)$; for example, the sequence

$$\sigma\langle 1,2,3,4, \ldots\rangle = \langle 2,3,4,5, \ldots\rangle.$$

$$\forall m \in \mathbf{N},\ \sigma^{-1}(y|[1,m]) = \{x \in \Pi : \forall n \in [2,m+1],\ x(n) = y(n-1)\},$$

which, according to 1.2, is open. Therefore, σ is a continuous function, called the *shift map* (on Π), but not to be confused with the map $r \to r+1$ also defined on $\mathbf{P}$. In this paper, all examples in this paper will concern the system $[\Pi;\sigma]$, and its subsystems.

The first study of the dynamics of (a variant of) $[\Pi;\sigma]$ appears in [16], but when $F \subseteq \mathbf{N}$ is finite, and in particular when F has just two elements (e.g., sequences of 0's amd 1's), the systems $[\Pi^{\mathbf{N}}F;\sigma]$ have been studied for more than 50 years (see [13] and [14] for more). They are called *symbolic dynamical systems* or *symbolic cascades*, and they have a variety of applications to Algebra and Number Theory (see [9]).

Using 1.2, the next lemma follows directly from the definitions.

2.1. Lemma. *For* $x \in \Pi$, *the following are true in* $[\Pi;\sigma]$:

1). x *is a fixed point iff* $\forall n \in \mathbf{N},\ x(n) = x(1)$.

2). x *is a periodic point iff* $\exists m \in \mathbf{N}\ \forall n \in \mathbf{N},\ x(n+m) = x(n)$ *iff* $\exists m \in \mathbf{N}$ *such that* $\forall k \in \mathbf{N},\ x|[km+1,(k+1)m] = x|[1,m]$ *as sequences.*

3). x *is almost-periodic iff* $\forall u \in \mathbf{N},\ \exists k = k_u \in \mathbf{N}$ *such that* $\forall m \in \mathbf{N}$,

$\{n \in \mathbf{N} : km \leq n < (k+1)m,\ \sigma^n x|[1,u] = x|[1,u]\} \neq \emptyset$

iff $\forall u \in \mathbf{N},\ \exists k = k_u \in \mathbf{N}$ *such that* $\forall m \in \mathbf{N}$,

$\{n \in \mathbf{N} : km \leq n < km+k,\ x|[n+1,n+u] = x|[1,u]\} \neq \emptyset$ *as sequences.*

4). x *is recurrent in iff* $\forall u \in \mathbf{N}$, $\exists a \geq u$ *such that* $\sigma^a x|[1,u] = x|[1,u]$ *iff* $\forall u \in \mathbf{N}$, $\exists a \geq u$ *such that* $x|[a+1,a+u] = x|[1,u]$ *as sequences.* \\//

Most fundamental to our constructions is the unique element in Π, denoted by α, given by

2.2. $\alpha(n) = k+1$ if k is the largest integer such that 2^k divides n.

The first few terms of α are <1,2,1,3,1,2,1,4,1,2,1,3,1,2,1,5, ...>.

The next lemma has a straight-forward proof by induction.

2.3. Lemma. *Suppose* $m,k \in \mathbf{N}$. *Then the following hold:*
1). *If* $n < 2^m$, *then* $\alpha(k2^m+n) = \alpha(n)$.
2). *If* $k \leq i < j \leq k+m$, *then* $\min\{\alpha(i),\alpha(j)\} \leq 1+\log_2 m$. \\//

2.4. Exercise. *Define points* $x,y \in \Pi$ *by* $\forall n \in \mathbf{N}$, $x(n) = n$, *and by* $y(n) = k$ *if* $n = 2^k$ *and* 1 *otherwise.*
1). *Prove that* x *and* y *are not recurrent.*
2). *Prove that* **OC**(x) *is countably infinite.*
3). *Prove that* **OC**(y) *contains the fixed point* <1,1,1,...>.
4). *Prove or disprove* **OC**(y) *is countably infinite.*

2.5. Example. *There is an almost-periodic point in* $[\Pi;\sigma]$ *which is not periodic.*

\\ Consider α. For $u \in \mathbf{N}$, let $k = 2^i$, where $i = \min\{j : u \leq 2^j\}$. Then 2.1(3) and 2.3(1) prove α is almost periodic. Notice that 2.1(2) shows that any periodic point in $[\Pi;\sigma]$ has finite range. However, $\alpha(2^m) = m+1$, and so α is not periodic. //

2.6. Exercise.
1). *Define* $x \in \Pi$ *by* $x(1) = 1$ *and, recursively, if* $x|[1,2^{n-1}]$ *has been defined, then for* m, $2^{n-1} < m \leq 2^n$, *define* $x(m) = 2$ *if* $x(m-2^{n-1}) = 1$; *otherwise define* $x(m) = 1$. So the first few terms of x are <1,2,2,1,2,1,1,2,2,1,1,2,1,2,2,1,...>. *Prove* x *is almost-periodic.*
2). *Construct a point* $x \in$ **OC**(α)**Orb**(α). (Hint. Consider <1,2,1,4;1,2,1,3,1,2,1,6, 1,2,1,3,1,2,1,4,1,2,1,3,1,2,1,5,1,2,1,3,1,2,1,4,1,2,1,3,1,2,1,8,...>.)
3). *A point* $p \in \Pi$ *is said to be* <u>*non-wandering*</u> *in* $[\Pi;\sigma]$ *provided* $\forall n,m \in \mathbf{N}$, $\exists x \in \Pi$ *with* $p|[1,m] = x|[1,m] = x|[n+1,n+m]$ *considered as sequences. Prove that each point of* $[\Pi;\sigma]$ *is non-wandering.*

The following result is well-known (with the same proof) in the case of the system $[\Pi^{\mathbf{N}}\{0,1\};\sigma]$.

2.7. Proposition. $[\Pi;\sigma]$ *is a transitive system.*

\\ Given $n \in \mathbf{N}$, the set $\mathcal{S}_n$ of all sequences of length n is the same as the set of all n-tuples consisting of elements of $\mathbf{N}$. Then $\mathcal{S}_n$ is countable. Since countable unions of countable sets is countable, the set $\mathcal{S}$ of all finite sequences in $\mathbf{N}$ is countable. Let $\{S_n : n\in \mathbf{N}\}$ list the elements of $\mathcal{S}$. Define a point $x\in \Pi$, recursively, by starting with S_1 and adjoining the sequence S_2 at its end. If we have the first m sequences adjoined in this manner, then adjoin S_{m+1} at the end. x will be the point constructed in this fashion. Given any point $y\in \Pi$ and $u\in \mathbf{N}$, then $y|[1,u]$ can be considered as a finite sequence. Thus, we may find $a\in \mathbf{N}$ such that $x|[a+1,a+u] = y|[1,u]$. Clearly, $y\in \mathbf{OC}(x)$. Therefore, $[\Pi;\sigma]$ is transitive. //

2.8. Example. *There is a recurrent point in* $[\Pi;\sigma]$ *which is not almost-periodic.*

\\ The point we use is the point x in the proof of 2.7. Notice that there are arbitrarily long constant finite sequences of the form $1+x(1)$. So for any $k\in \mathbf{N}$, we may find $a,b\in \mathbf{N}$ such that $b > k$ and $\forall n\in [a,a+b]$, $x(n) \neq x(1)$. Therefore, x is not almost-periodic. //

3. More homeomorphs of P

We will build new dynamical systems on $\mathbf{P}$ using two major tools: Important is a theorem, 3.2, considerably expanding 1.3 and its consequence, "technology" expanding the essentional idea in the proof of 2.7. The first theorem of this section is completely new and shows why $\mathbf{P}$ is so important to Topological Dynamics. It says, "Any odd behavior of a discrete dynamical system on a separable complete metrizable space should be reflected by a system on $\mathbf{P}$."

3.1. Theorem. *Suppose* $[X,f]$ *is a system on a separable complete metrizable space* (X,d). *Then there is a system* $[\mathbf{P},g]$ *and a homomorphism* $h : [\mathbf{P},g] \to [X,f]$.

\\ $\forall n\in \mathbf{N}$, and $\forall r\in {}^{[1,n]}\mathbf{N}$, we will define, recursively, an open set G_r in X, and, if $n > 1$, and an $r^*\in {}^{[1,n-1]}\mathbf{N}$ all subject to the following conditions:

1). $\{G_r : r\in {}^{[1,1]}\mathbf{N}\}$ is an open cover of X.
2). If $n > 1$ and if $r\in {}^{[1,n-1]}\mathbf{N}$, then $G_r = \cup\{G_s : s\in {}^{[1,n]}\mathbf{N}, r = s|[1,n-1]\}$.
3). If $n > 1$, then the diameter $\delta(G_r)$ of G_r, is at most $\dfrac{\delta(G_{r|[1,n-1]})}{2}$.
4). If $n > 1$ and if $r\in {}^{[1,n]}\mathbf{N}$, then $cl(G_r) \subseteq G_{r|[1,n-1]}$.
5). If $n > 1$ and if $r\in {}^{[1,n]}\mathbf{N}$, then $f(G_r) \subseteq G_{r^*}$.
6). If $n > 2$ and if $r\in {}^{[1,n]}\mathbf{N}$, then $G_{r|[1,n-1]^*} = G_{r^*|[1,n-2]}$.

Since X is Lindelöf, there is a countable open cover $\mathcal{R}$ of X by sets of diameter 1. Allowing repeats (in the case $\mathcal{R}$ is finite), let $\{R_m : m\in \mathbf{N}\}$ index $\mathcal{R}$. $\forall r\in {}^{[1,1]}\mathbf{N}$, define $G_r = R_{r(1)}$. Thus, for $n = 1$, the conditions (1)-(6) are satisfied.

Suppose $m > 1$ and $\forall n < m$, $\forall r\in {}^{[1,n]}\mathrm{N}$, we have defined G_r, and if $n > 1$, we have defined G_{r^*} to satisfy the conditions (1)-(6). Fix $r\in {}^{[1,m-1]}\mathbf{N}$. $\forall x\in G_r$,

(2) finds $t(x) \in {}^{[1,m-1]}\mathbf{N}$ such that $f(x) \in G_{t(x)}$. So choose an open nhbd $U_x \subseteq G_r$ of x such that $f(U_x) \subseteq G_{t(x)}$. In case, $m > 2$, then, by recursion and (5), $f(G_r) \subseteq G_{r^*}$. Again using (2),

$$G_{r^*} = \cup\{G_s : s \in {}^{[1,m-1]}\mathbf{N},\ r^* = s|[1,m-1]\}.$$

When $n > 2$, we may choose $t(x)$ such that $r^* = t(x)|[1,m-2]$. Now choose an open ball $B(x,\varepsilon_x)$ centered at x with $B(x,\varepsilon_x) \subseteq U_x$. As X is separable metric, each subspace of X is Lindelöf. Thus, we may choose a countable subset $\{V_k : k \in \mathbf{N}\}$ of $\{B(x,\frac{\varepsilon_x}{2}) : x \in G_r\}$ covering G_r. $\forall t \in {}^{[1,m]}\mathbf{N}$, with $r = t|[1,m-1]$, let $G_t = V_{t(m)}$. Clearly, the conditions (1)-(6) are satisfied.

We may now assume the construction of the sets G_r and functions r^* are defined $\forall n \in \mathbf{N}$, and $\forall r \in {}^{[1,n]}\mathbf{N}$, $\forall r^* \in {}^{[1,n-1]}\mathbf{N}$. Suppose $\pi \in \Pi$. According to (2) and (3), $\{cl(G_{\pi|[1,n]}) : n \in \mathbf{N}\}$ is a descending family of closed sets with diameters converging to 0. Since X is complete, there is a unique element $x_\pi \in \cap_{n \in \mathbf{N}} cl(G_{\pi|[1,n]})$. Define a function $h : \Pi \to X$ by $h(\pi) = x_\pi$. Notice that given $x \in X$, (1) and (2) inductively define $\pi \in \Pi$ such that $x \in \cap_{n \in \mathbf{N}} G_{\pi|[1,n]}$. Thus, h is surjective.

We show h is continuous. Suppose $\pi \in \Pi$ and $h(\pi) = x \in X$. By (4),

$$\cap_{n \in \mathbf{N}} cl(G_{\pi|[1,n]}) = \cap_{n \in \mathbf{N}} G_{\pi|[1,n]}.$$

(3) shows $\{G_{\pi|[1,n]} : n \in \mathbf{N}\}$ is a nhbd base at x. So $h(\theta) = x_\theta \in G_{\pi|[1,n]}$ implies $\exists m > n$ with $G_{\theta|[1,m]} \subseteq G_{\pi|[1,n]}$. Thus, $h(\{\sigma \in \Pi : \sigma|[1,n] = \theta|[1,n]\}) \subseteq G_{\pi|[1,n]}$. Therefore, h is continuous.

Now consider $\pi \in \Pi$. According to (3), (4), (5), and (6), there is a unique $\pi^* \in \Pi$ such that $\forall n \in \mathbf{N}$, $\pi^*|[1,n] = (\pi|[1,n])^*$. So we may define a function $g : \Pi \to \Pi$ by $g(\pi) = \pi^*$. Let $\theta \in \Pi$ have $\theta^*|[1,m] = \pi^*|[1,m]$. Clearly,

$$g(\{\sigma \in \Pi : \sigma|[1,m+1] = \theta|[1,m+1]\}) \subseteq \{\tau \in \Pi : \tau|[1,m] = \pi^*|[1,m]\}).$$

Therefore, h is continuous.

Fix $\pi \in \Pi$. Then $fh(\pi) = f(x_\pi) \in \cap_{n \in \mathbf{N}} G_{\pi^*|[1,n]} = \{x_{\pi^*}\}$. But $hg(\pi) = x_{\pi^*}$. Therefore, h is a homomorphism. //

Though the following result was probably discovered 50 or more years ago (I have no reference), it was unpublished folklore until its first appearance in **[16]**. An expansion of that proof was used to prove 3.1.

3.2. Corollary. *A space* X *is homeomorphic to* $\mathbf{P}$ *iff it is a separable zero-dimensional completely metrizable nowhere-locally compact space.*

\\. <u>ONLY IF</u>. Recall $\mathbf{Q}$ is the set rationals. Fix $p \in \mathbf{P}$, and let $Q = \{p+q : q \in \mathbf{Q}\}$. Between each pair of reals is a member of $\mathbf{Q}$, so given $r,s \in \mathbf{P}$, $\exists q \in \mathbf{Q}$ such that q is between r-p and s-p. But $p+q$ is between r and s; thus, Q witnesses that $\mathbf{P}$, and any homeomorphs, is separable. As any open set of $\mathbf{P}$ contains a sequence converging to a missing rational, each compact set in **P** has empty interior; hence, **P** is nowhere-locally compact. **P** is not complete, but it is, by 1.7, homeomorphic to Π which is a complete mtric space - see any standard topology text; e.g., [7]. Thus, **P**, and any of its homeomorphs, is completely metrizable.

IF. Assume X is a separable zero-dimensional nowhere locally compact complete metric space. Then X has a base $\mathfrak{C}$ consisting of clopen sets. Given $C \in \mathfrak{R}$, C is complete since it is closed in X. C is not compact since it is open and all compact sets in X have empty interior. However, C is Lindelöf; hence, it is the union a countably infinite family of pairwise disjoint clopen subset $\mathcal{S}(C) \subseteq \mathfrak{C}$ of sets with diameter at most half the diameter of C. Thus, within the proof of 3.1, we can require an extra condition:

7). If $r,s \in {}^{[1,n]}\mathbf{N}$, $r \neq s$, and $r = s|[1,\text{dom}(r)-1] = s|[1,\text{dom}(s)-1]$, then $G_s \cap G_t = \varnothing$.

Thus, the function $h : \Pi \to X$ is an injection. Given $r \in {}^{[1,n]}\mathbf{N}$. Then (7) also implies $h(\{\pi \in \Pi : \pi|[1,n] = r\}) = G_r$. Therefore, h is a homeomorphism. //

3.3. Lemma. *A closed subset* A *of* Π *is compact iff it is bounded in the pointwise product partial order.*

\\ IF. Suppose A is a bounded closed subset A of Π; i.e, there is $f \in \Pi$ such that $A \subseteq K = \Pi_{n \in \mathbf{N}} [1,f(n)]$. As K is the product of finite sets, K is compact. Since A is a closed subset of K, it is compact.

ONLY IF. Suppose K is a compact set in Π. For $m \in \mathbf{N}$, K projects on to the mth-coordinate as a compact, and hence finite subset of **N** with maximum f(n). As $K \subseteq \Pi_{n \in \mathbf{N}} [1,f(n)]$, A is bounded. //

Here is the tool we spoke about at the beginning of this section.

3.4. Theorem. *A non-empty closed subspace* X *of* Π *is homeomorphic to* **P** *iff each non-empty open set of* X *is unbounded in the pointwise product partial order.*

\\ Since, from 3.2, X is nowhere-locally compact, 3.3 immediately implies the properties above. Conversely, suppose $X \subseteq \Pi$ has the properties above. Clearly, each subspace of Π is zero-dimensional separable metric, so X is separable. 3.3 shows X is nowhere locally compact iff each of its non-empty open subsets are unbounded. Therefore, 3.2 applies to prove the result. //

4. Constructing points in Π

Here we introduce our second tool used to build and discuss points in Π, we need some special notions about finite sequences in **N**. A *block* will be a finite function with domain dom B an (a possibly empty) interval in **N**, and range rng B a subset of **N**. All blocks will be assumed to have as domain an initial segment of **N** unless otherwise stated. If the block B has (non-) empty domain, we write $B = \varnothing$ $(B \neq \varnothing)$. The length of a block B is $\lambda B = \text{card}(\text{dom } B)$. If $B \neq \varnothing$ is a block, if $a \leq b \in \mathbf{N}$, and if dom B = [a,b], then by a *tail* (a *head*) of B, we mean any block of the form $B|[c,b]$ (respectively, $B|[a,c]$), where $a \leq c \leq b$.

4.1. A partial order on blocks. Suppose A and B are blocks. We will say A is a *copy* of B and write $A \equiv B$ provided the following two conditions are satisfied:

1). $\lambda A = \lambda B$ and

2). $\exists z \in \mathbb{N}$ such that $\forall n \in \text{dom } A$, $A(n) = B(n+z)$.

So $\equiv$ is an equivalence relation on the set of all blocks. We will also need a partial order on the equivalence classes: Let us agree that A is *in* B and write $A \leq B$ provided there is an interval $I \subseteq \text{dom } B$ such that $A \equiv B|I$.

We can expand these ideas to points in Π. For $x \in \Pi$, we say A is a block in x and we write $A \leq x$ provided there is an interval $I \subseteq \mathbf{N}$ such that $A \equiv x|I$. Suppose B is a block and $I \subseteq \mathbf{N}$ is an interval. We say that $x|I$ is a *maximal tail* of B in x provided $x|I$ is a tail of B and for each interval $J \supsetneq I$, $x|J \nleq B$.

4.2. A finitary operation on classes of blocks. Now suppose I and J are finite intervals in $\mathbf{N}$, f is a function with dom f = I and rng f = J. Further, suppose $(B_z : z \in J)$ is a sequence of blocks. Let $f(B_z : z \in I)$ or $(B_{fz} : z \in J)$ denote the unique (up to the equivalence $\equiv$) block obtained by allowing $\forall z$, $\min I \leq z < \max I$, B_{fz} to be immediately followed by $B_{f(z+1)}$. When $f : [a,b] \to [a,b]$ is the identity function, we also let $B_a \wedge B_{a+1} \wedge \cdots \wedge B_b$ denote $f(B_z : z \in [1,n])$.

4.3. Exercise. *Suppose* $n \in \mathbf{N}$ and $(B_z : z < 2^n)$ *is a family of blocks. Prove that*

$$\lambda\alpha(B_z : z < 2^n) = \sum_{k=1}^{n} 2^{n-k} \lambda B_k .$$

4.4. Lemma. *For* $x \in \Pi$, $\mathbf{OC}(x)$ is homeomorphic to $\mathbf{P}$ *provided that for each block* A *in* x, *there is an* $m \in \mathbf{N}$ *such that*

$$L = \{\text{rng } C : \lambda C = m \text{ and } A \wedge C \text{ is a block in } x\} \text{ is infinite.}$$

\\\\ Given $a < b \in \mathbf{N}$ let $G_{a,b} = \{y \in \mathbf{Orb}(x) : y|[1,b-a] \equiv x|[a,b]\}$. Then $G_{a,b}$ is open in $\mathbf{OC}(x)$ and each open set in $\mathbf{OC}(x)$ contains a $G_{a,b}$ for suitable $a < b \in \mathbb{N}$. Let

$$\mathfrak{C} = \{\text{blocks } C : \lambda C = m \text{ and } A \wedge C \text{ is a block in } x\}.$$

Since L is infinite, there is a first $k \leq m$ such that $\{C(k) : C \in \mathfrak{C}\}$ is unbounded. As $\forall C \in \mathfrak{C}, \lambda C = m$, $\{C|[1,k) : C \in \mathfrak{C}\}$ is finite. Without loss of generality, we may assume that $\text{card}(\{C|[1,k) : C \in \mathfrak{C}\}) = 1$ and $\forall C \in \mathfrak{C}$, $C = C|[1,k]$. Suppose $C \in \mathfrak{C}$. Let $A = x|[u,v]$ and $w = v+k$. Then $G_{u,w} \subseteq G_{u,v}$. But if $A = x|[a,b]$, then $G_{u,v} = G_{a,b}$ and $\{y(b-a+k) : y \in G_{a,b}\}$ is unbounded. Hence, $G_{a,b}$ is unbounded. According to 3.4, $\mathbf{OC}(x)$ is homeomorphic to $\mathbf{P}$. //

4.5. Corollary. *For* $x \in \Pi$, $\mathbf{OC}(\alpha)$ *is homeomorphic to* $\mathbf{P}$.

\\ According to 2.3(1), $\forall m \in \mathbf{N}$, $\alpha|[1,2^m) \equiv \alpha|[2^m,2^{m+1})$. Since $\alpha(2^{m+1}) = m$, the hypothesis of 4.4 is satisfied. //

A *street* is a sequence $S = (B_z : z \in \mathbf{N})$ of blocks indexed by $\mathbf{N}$ and such that $\forall z \in \mathbf{N}$, $B_z \neq \emptyset$.

4.6. An infinitary operation on streets. Suppose $S = (B_z : z \in \mathbf{N})$ is a street and $x \in \Pi$. Then xS denotes the unique element y of Π such that $\forall m \in \mathbf{N}$, if $f = x|[1,m)$, and if $s = \sum_{z=1}^{m-1} \lambda B_{fz}$, then $y|[1,s] = f(B_z : z \in [1,m))$.

4.7. Lemma. *Suppose* $S = (B_z : z \in \mathbf{N})$ *is a street, suppose* $f \in \Pi$ *is an increasing function, and suppose* $k \in \mathbf{N}$ *satisfies*

$$I = \{z \in \mathbf{N} : k \leq \lambda B_z, B_z(k) \geq f(z)\}$$

infinite. Then $\mathbf{OC}(\alpha S)$ *is homeomorphic to* $\mathbf{P}$.

\\ Suppose A is a block in $x = \alpha S$. Let $i \in I$ be such that $A \leq \alpha(B_w : w \in [1,2^i))$, say $A = x|[a,b]$ for

$$1 \leq a \leq b < s = \sum_{w=1}^{2^i} \lambda B_{\alpha(w)} .$$

$\forall z \in I$ with $z \geq i$, let $C_z = x|[s-b,s] \wedge B_z|[1,k]$. From 2.3,

$$\{\operatorname{rng} C_z : z \in I, A \wedge C_z \text{ is a block in } x\}$$

is infinite. For $m = k+s-b$, the hypothesis of 4.4 is satisfied. Therefore, $\mathbf{OC}(\alpha S)$ is homeomorphic to $\mathbf{P}$. //

5. Recurrent points in $[\Pi;\sigma]$

5.1. Lemma. *Suppose* ϕ *is recurrent in* $[\Pi;\sigma]$ *and* S *is a street. Then* ϕS *is recurrent in* $[\Pi;\sigma]$.

\\ Suppose $x = \phi S$. Fix $m \in \mathbf{N}$ and let $p > m$ be such that $x|[1,m]$ is a head of $\phi(B_z : z \leq p)$. As ϕ is recurrent, 2.1(4) finds $k > p$ with $\phi|[k+1,k+p] \equiv \phi|[1,p]$. So

$$\phi(B_z : k+1 \leq z \leq k+p) \equiv \phi(B_z : z \leq p).$$

Thus, $x|[1,m]$ is a head of $\phi(B_z : k+1 \leq z \leq k+p)$. From 2.1(4), x is recurrent. //

5.2. Theorem. *Suppose* $x \in \prod$. *Then* x *is recurrent in* $[\prod;\sigma]$ *iff there is a street* S *such that* $x = \alpha S$.

\\ The "if" is a consequence of 5.1 once we see α is recurrent in $[\prod;\sigma]$. But 2.3(1) shows α is almost-periodic!

Only if. Conversely, suppose x is recurrent in $[\prod;\sigma]$. By recursion, we construct a street $(B_z : z \in \mathbf{N})$. Let $B_1 = x|\{1\}$, and suppose $m \in \mathbf{N}$ is such that $\forall z \leq m$, the following hold:

1). B_z is defined (and hence, $\forall z \in [1,2^m)$, $B_{\alpha z}$ is defined).

2). $x|[1,s] \equiv f(B_z : z \in [1,2^m))$, where $s = \sum_{z=1}^{2^m-1} \lambda B_{\alpha z}$, where $f = \alpha|[1,2^m)$.

Since x is recurrent, we may find a smallest $b > s$ such that $x|[1,s] \equiv x|[b+1,b+s]$. Now define $B_{m+1} = x|[s+1,b]$. As it is clear that we can proceed, in the above fashion, defining B_z and exhausting x, condition (2) shows $x = \alpha(B_z : z \in \mathbf{N})$. //

Unlike 5.2, for recurrence, **I know of no interesting characterization for almost periodic points in** $[\prod;\sigma]$. However, 5.3 and 5.4 below exhibit what is known at present.

5.3. Theorem. *Suppose* $S = (B_z : z \in \mathbf{N})$ *is a street.*

1). *If* $\{\lambda B_z : z \in \mathbf{N}\}$ *is finite and if* ϕ *is almost-periodic in* $[\prod;\sigma]$, *then* ϕS *is almost-periodic in* $[\prod;\sigma]$.

2). *If* $x = \phi S$ *is almost-periodic in* $[\prod;\sigma]$ *and if* $\exists m \in \mathbf{N}$ *such that*
$$Z = \{z \in \mathbf{N} : \alpha(B_w : w \in [1,2^{m+1})) \text{ is not a block in } B_z\}$$
is infinite, then $L = \{\lambda B_z : z \in I\}$ *is finite.*

\\ 1). Let $x = \phi S$ and suppose $u \in \mathbf{N}$ is an upper bound for $\{\lambda B_z : z \in \mathbf{N}\}$. Fix $m \in \mathbf{N}$ and let $p > m$ such that $x|[1,m]$ is a head of $\phi(B_z : z \leq p)$. As ϕ is almost-periodic, 2.1(3) finds $k = k_p \in \mathbf{N}$ such that $\forall q \in \mathbf{N}$,

$$\{n \in [kq,kq+k) : \phi|[n+1,n+p] \equiv \phi|[1,p]\} \neq \emptyset.$$

Hence, $\forall q \in \mathbf{N}$,

$$\{n \in [kq,(k+1)q) : \phi(B_z : n+1 \leq z \leq n+p) \equiv \phi(B_z : z \leq p)\} \neq \emptyset.$$

Therefore, $\forall q \in \mathbf{N}$,

$$\{n \in [kuq,(ku+1)q) : \phi|[n+1,n+p] \equiv \phi|[1,p]\} \neq \emptyset,$$

and, according to 2.1(3), x is almost-periodic in $[\prod;\sigma]$.

2). Suppose L is infinite and $k \in \mathbf{N}$. We may choose $z \in Z$ such that

$$\lambda B_z \geq k + \lambda\alpha(B_w : w \in [1, 2^{m+1})).$$

But then B_z is a block in x of length greater than k failing to contain a copy of $\alpha(B_w : w \in [1, 2^{m+1}))$. From 2.1(3), x is not almost-periodic. //

5.4. Example. *There is an almost periodic-point* x *such that if* $(S_z : z \in \mathbf{N})$ *is a street with* $x = \alpha(S_z : z \in \mathbf{N})$. *Then* $\{\lambda S_z : z \in I\}$ *is infinite.*

\\ Define $B_1 = \langle 1 \rangle$, and for $n \geq 2$, define

1). $B_n = \langle n \rangle \wedge \alpha(B_k : k < 2^{n-1}) \wedge \langle n \rangle$.

Let $\beta = \alpha(B_n : n \in \mathbf{N})$. Then the first few terms of β are $\langle 1,2,1,2,1,3,1,2,1,2,1,3,1,2,1,2,1,4,1,2,1,2,1,3,1,2,1,2,1,3,1,2,1,2,1,4,1,2,1,2,1,3,1,2,1,2,1,3,1,2,1,2,1,5, \ldots \rangle$.
Clearly, 2.2, 4.3, and finite induction proves the next two statements

2). If $n > 1$, then $\lambda B_n = 2 \cdot 3^{n-1} + 1$ and $\lambda\alpha(B_k : k < 2^n) = 2 \cdot 3^{n-2} - 1$.
3). $\forall n, m \in \mathbf{N}$, $\alpha(B_k : k < 2^n) \leq \beta | [m+1, m+4 \cdot 3^{n-1} - 1]$.

Acording to 2.1(3), (3) shows β is almost periodic.

Let $t \in \mathbf{N}$ and let $(S_z : z \in \mathbf{N})$ be a street such that $\beta = \alpha(S_z : z \in \mathbf{N})$ and $\forall z \in \mathbf{N}$, $\lambda S_z \leq t$. Choose $m \in \mathbf{N}$ to be the first integer with $t \leq 2 \cdot 3^{m-1}$. Choose $r \in \mathbf{N}$ as the first integer with $m \in \operatorname{rng} S_r$. Let $R_1 = \alpha(S_z : z < 2^{r-1})$, and $A_1 = \alpha(B_n : n < 2^{m-1})$. Since $m \notin \operatorname{rng} R_1$, we have

4). R_1 is a head of A_1 which is a head of β.

So (4) shows

5). $A_1 \wedge \langle m \rangle$ is a head of $R_1 \wedge S_r$ which is a head of β.

From (1), $A_1 \wedge \langle m \rangle \wedge A_1$ and $A_1 \wedge \langle m \rangle \wedge A_1 \wedge \langle m \rangle$ are heads of β. As $\lambda S_r \leq t$, (2) shows $\langle m \rangle$ appears in S_r but once. Using 2.3, 4.6, and (4), we have

6). $R_1 \wedge S_r \wedge R_1$ is a head of $A_1 \wedge \langle m \rangle \wedge A_1$.

From (6) there is a second integer $s \in \mathbf{N}$ with $m \in S_s$. Let $R_2 = \alpha(S_z : z < s)$, and $A_2 = A_1 \wedge \langle m \rangle \wedge A_1 \wedge \langle m \rangle \wedge A_1$. Clearly,

7). $A_1 \wedge \langle m \rangle \wedge A_1 \wedge \langle m \rangle$ is a head of $R_1 \wedge S_r \wedge R_1 \wedge S_s$ which is a head of β.

Now (5) and (7) show $A_2 \wedge \langle m \rangle$ is a subsequence of $T = R_1 \wedge S_r \wedge R_1 \wedge S_s \wedge R_1 \wedge S_r \wedge R_1$. Since $\forall z \le s$, $\lambda S_z \le t$, rng S_z is bounded by m. Since T is a head of β, T must be a head of A_2. Thus, $\langle m,m,m \rangle$ is a subsequence of A_2 which contradicts (1). //

5.5. Lemma. *Suppose* $S = (B_z : z \in \mathbf{N})$ *is a street,* $y \in \mathbf{OC}(\alpha S)$, *and suppose* $\exists \lambda \in \mathbf{N}$ *such that for each block* F *with* $\lambda F = \lambda$, $\{z \in \mathbf{N} : F \text{ is a head or tail of } B_z\}$ *is finite. If* y *is not recurrent in* $[\Pi;\sigma]$ *or if* $\mathbf{OC}(y) \ne \mathbf{OC}(\alpha S)$, *then* $\exists h \in \Pi$ *such that* $\forall z \in \mathbf{N}$, $y|[1,z] \le B_{hz}$.

\\ Let $x = \alpha S$. For simplicity, we will use α even when we restrict its domain. Suppose $m \in \mathbf{N}$, we define h(m).

Since $y \in \mathbf{OC}(x)$, $\forall m \in \mathbf{N}$, $\exists k \in \mathbf{N}$ $y|[1,m] = \sigma^k x|[1,m]$, so there are functions $a,b : \mathbf{N} \to \mathbf{N}$ such that

1). $a(m) \le b(m)$, and
2). $\forall m \in \mathbf{N}$, $b(m)-a(m)$ is minimal with respect to $(y|[1,m]) \le \alpha(B_z : a(m) \le z \le b(m))$.

Now, $\forall m \in \mathbf{N}$, find $c(m) \in \mathbf{N}$ and (possibly empty) blocks H(m) and T(m) such that

3). $H(m) \wedge (y|[1,m])$ is a head of $\alpha(B_z : a(m) \le z \le c(m))$,
4). T(m) is a tail of $\alpha(B_z : c(m) \le z \le b(m))$, and
5). $a(m) \le c(m) \le b(m)$ and $H(m) \wedge (y|[1,m]) \wedge T(m) \equiv \alpha(B_z : a(m) \le z \le b(m))$.

Notice that the conditions (2) to (5) implies

6). $\alpha\big(B_z : a(m) < z < b(m)\big) \le (y|[1,m])$.

From (6), $\exists f,g : \mathbf{N} \to \mathbf{N}$ such that $\forall m \in \mathbf{N}$, $(y|[f(m),g(m)]) \equiv B_{\alpha c(m)}$. According to the hypothesis of this lemma (concerning heads and tails of B_z), we have one of two possibilities:

7). Either there is an infinite $N \subseteq \mathbf{N}$ such that both $f|N$ and $g|N$ are monotone, or
8). there is an infinite $N \subseteq \mathbf{N}$ such that $(\alpha c)|N$ is constant.

Of course if (7) holds, we are done - just set $h(m) = \alpha c(i_m)$, where the natural indexing of N is $\{i_m : m \in \mathbf{N}\}$. So we assume (8) is true. From the hypothesis, we have just two possibilities for the functions c−a and b−c.

CASE1. *For each infinite* $I \subseteq N$, *both* rng $(c-a)|I$ *and* rng $(b-c)|I$ *are unbounded.*
CASE2. *There is an infinite* $I \subseteq N$ *such that at least one of* rng $(c-a)|I$ *and* rng $(b-c)|I$ *is bounded, while the other is unbounded.*

Assume CASE1 holds, and suppose $k \in \mathbf{N}$ is arbitrary. As $y \in \mathbf{OC}(x)$, $\exists p \in \mathbf{N}$ such that $x|[p+1,p+k] = y|[1,k]$. Thus, we may find $k_1 \in \mathbf{N}$ such that

9). both $x|[1,k]$ and $y|[1,k]$ are in $\alpha(B_z : z \leq k_1)$.

Let $I = \{m \in \mathbf{N} : 2^{k_1+1}+1 < c(m)-a(m)\}$. Since CASE1 holds, I is infinite. Again applying CASE1 shows $J = \{m \in I : 2^{k_1+1}+1 < b(m)-c(m)\}$ is infinite. Now choose $j \in J$. Using (6), 2.3(1) shows

10). $\alpha(B_z : z \leq k_1) \leq y|[1,j]$.

Applying (6) to (10), we see $x|[1,k]$ and $y|[1,k]$ are in $y|[1,j]$. It is now clear that $\exists q,r \in \mathbf{N}$ such that $\sigma^q y|[1,k] = x|[1,k]$, and $\sigma^r y|[1,k] = y|[1,k]$. As $k \in \mathbf{N}$ is arbitrary, we see, respectively, that $\mathbf{OC}(y) = \mathbf{OC}(x)$ and so y is recurrent.

Assume CASE2 is true - say rng $(c-a)|I$ is bounded. Then, without loss of generality, we may assume:

11). $(c-a)|I$ is constant, $\exists c_0 \in \mathbf{N}$ such that $\forall m \in I$ $c(m) = a(m)+c_0$.

From (8), (11), and 2.3, $\exists j \in [0,c_0]$ such that $\forall i \in [0,c_0)\backslash\{j\}$, $\alpha(c+i)|(I \cap [k,\infty))$ is constant for $k = 1+\log_2(\sum_{z=1}^{c_0} \lambda B_z)$. From (3), $j = 0$. Hence, $\exists b_0 \in \mathbf{N}$ such that $\forall m > k$, $m \in I$, $y|[1,b_0] \equiv \alpha(B_z : a(m) < z \leq c(m))$. Therefore, there is an increasing $d : I \to \mathbf{N}$ and an infinite valued $e : I \to \mathbf{N}$ such that $(y|[-d(m),a_0))$ is a tail of $B_{\alpha ce(m)}$ - a contradiction.

The proof, in case rng $(b-c)|I$ is bounded, is similar. //

5.6. Lemma. *Suppose* $S = (B_z : z \in \mathbf{N})$ *is a street satisfying*

$$\#).\ \forall n < m \in \mathbf{N},\ \text{rng}\, B_n \cap \text{rng}\, B_m = \varnothing .$$

Then $\mathbf{OC}(\alpha S)$ *is minimal, and* $y \in \mathbf{OC}(\alpha S)$ *iff* y *can be written as*

$$(T \wedge \alpha(B_z : z < h(1))) \wedge \bigwedge_{n>1} (B_{\alpha(h(n))} \wedge \alpha(B_z : z \in [1,h(n))))),$$

where $h \in \Pi$ *is increasing and* T *is a maximal tail of* $B_{\alpha(h(1))}$ *in* y.

\\. To see that $[\mathbf{OC}(\alpha S);\sigma]$ is minimal, suppose $x \in \mathbf{OC}(\alpha S)$. As the B_z's are pairwise-disjoint, the hypothesis of 5.5 is satisfied. So if $\mathbf{OC}(x) \neq \mathbf{OC}(\alpha S)$, then 5.5 finds an $h \in \Pi$ such that $\forall n \in \mathbf{N}$, $x|[1,n] \leq B_{h(n)}$. Clearly, h is constant, say $h(1) = m$. Choose $n > 2^m$, then $n = \lambda x|[1,n] \leq \lambda B_m \leq 2^m$ - ridiculous. Thus, $\mathbf{OC}(x) = \mathbf{OC}(\alpha S)$. Therefore, $[\mathbf{OC}(\alpha S);\sigma]$ is minimal.

IF Just check (see exercise 2.6(2), where $T = \langle 2 \rangle$) that $\exists k \geq 0$ such that

$$\sigma^k(\alpha S) = (T\wedge\alpha(B_z : z < h(1)))\wedge(\wedge_{1<n<m}(B_{\alpha(h(n))}\wedge\alpha(B_z : z\in[1,h(n))))) \wedge$$
$$\wedge B_{\alpha(h(m))}\wedge\alpha(B_z : z > m).$$

ONLY IF. Suppose $y\in \mathbf{OC}(\alpha S)$. Then $\exists a\in \mathbf{Orb}(\alpha S)$ and $\exists z,n\in \mathbf{N}$ such that $B_z(n) = a(1) = y(1)$. Choose the first $h(1)\in \mathbf{N}$ such that $\alpha(h(1)) = z$. Let

$$p = \lambda B_z\text{-}(n\text{-}1)+\lambda\alpha(B_k : k < h(1)).$$

From (#), if $x\in \mathbf{Orb}(\alpha S)$ and if $x(1) = B_z(n)$, then $x|[1,\lambda B_z\text{-}(n\text{-}1)] = B_z|[n,\lambda B_z]$. WLOG we may assume $a|[1,p] = y|[1,p]$, there is a tail T of B_z which is a head of y. WLOG we may assume T is a maximal tail of B_z in y. Further, (#) and 2.3(1) imply that whenever $x\in \mathbf{Orb}(\alpha S)$ and T is a head of x, then $T\wedge\alpha(B_k : k < h(1))$ is a head of x. So $T\wedge\alpha(B_k : k < h(1)) = a|[1,p]$ is a head of y.

Suppose $m\in \mathbf{N}$ and $\forall n < m$ we have defined $h(n)$ so that $h|[1,m)$ is an increasing sequence and

*). $H = (T\wedge\alpha(B_z : z < h(1)))\wedge\wedge_{1<n<m}(B_{\alpha(h(n))}\wedge\alpha(B_z : z\in[1,h(n))))$ is a head of y.

Let $q = 1+\lambda H$. Then $\exists b\in \mathbf{Orb}(\alpha S)$ with $b|[1,q] = y|[1,q]$. Find the first $u\in \mathbf{N}$ with $\langle b(q)\rangle \leq B_{\alpha(u)}$. As $\langle\alpha(h(m-1))\rangle\wedge\alpha|[1,h(m-1))$ is a block in α, 2.2, (#), and (*) imply $u > h(m-1)$. Let $h(m) = u$. Since $b\in \mathbf{Orb}(\alpha S)$, we can again apply (#) to show $b(q) = B_{\alpha(u)}(1)$. Now follow the methods of the previous paragraph to show, in succession, the following are heads of y:

$$(T\wedge\alpha(B_z : z < h(1)))\wedge\wedge_{1<n<m}(B_{\alpha(h(n))}\wedge\alpha(B_z : z\in[1,h(n))))\wedge B_{\alpha(u)}$$
$$(T\wedge\alpha(B_z : z < h(1)))\wedge\wedge_{1<n\leq m}(B_{\alpha(h(n))}\wedge\alpha(B_z : z\in[1,h(n))))$$

Thus, the construction of h is completed by recursion, and y clearly satisfies the conclusion. //

6. Systems on P with all points recurrent.

We begin with two standard and easy to prove results in Topological Dynamics. Note 6.1 is usually stated for the compact case, but x is remains almost-perodic in $[\beta X,\beta f]$, where βX is the Stone-Čech compactification of X and βf is the extension of f to βX.

6.1. *If* x *is almost-periodic in a system* $[X;f]$, *then* $\mathbf{OC}(x)$ *is minimal in* $[X;f]$ *and each point of* $\mathbf{OC}(x)$ *is almost-periodic.*

6.2. *If* $[X;f]$ *is a minimal system, then each point of* $[X;f]$ *is recurrent.*

Our next example was our first example and lead to the discovery of α. Please contrast it with 6.1.

6.3. Example. *There is a continuous function* f: $\mathbf{P} \to \mathbf{P}$ *such that* $[\mathbf{P};f]$ *is a minimal system, and no point of* $\mathbf{P}$ *is almost-periodic in* $[\mathbf{P};f]$.

\\\\. $\forall z \in \mathbf{N}$, let B_z be the constant block z of length z. Let $x = \alpha(B_z : z \in \mathbf{N})$. According to 4.7, $\mathbf{OC}(x)$ is homeomorphic to $\mathbf{P}$. 5.6 shows $[\mathbf{OC}(x);\sigma]$ is minimal.

Aplying 6.1 it is sufficient to prove x is not almost-periodic. According to 1.2, $U = \{y \in \Pi : y(1) = 1\}$ is a neighborhood of x. Of course, $\sigma^n x \in U$ iff $x(n+1) = 1$ iff $z \neq 1$ implies $x(n+1) \notin \text{rng } B_z$. Fix $m \in \mathbf{N}$. $\forall z \geq m$, dom B_z has length z and $1 \notin \text{rng } B_z$. Therefore, 2.1(4) shows x cannot be almost-periodic. //

The next example is an expansion of the idea of proof of 5.4 and should be contrasted with 6.2.

6.4. Example. *There is a continuous function* $f : \mathbf{P} \to \mathbf{P}$ *such that each point in* $\mathbf{P}$ *is recurrent in* $[\mathbf{P};f]$, *while* $[\mathbf{P};f]$ *has no minimal sets.*

\\\\. $\forall n \in \mathbf{N}$, define $I_n = \{2^{m-1}(2n-1) : m \in \mathbf{N}\}$. During the remainder of this proof we will consider blocks as either functions or ordered sequences. $\forall n \in \mathbf{N}$, let B^n_1 denote the one-element sequence $<2n-1>$, and define, recursively

1). $\forall m \in \mathbf{N}$, $\forall n > 1$, $m \neq 0$,

$$B^n_m = <2^{m-1}(2n-1)> \wedge \alpha(B^{n+1}_z : z < 2^{m-1}) \wedge <2^{m-1}(2n-1)>.$$

Here are samples: $B^1_1 = <1>$, $B^2_1 = <3>$, $B^1_2 = <2,3,2>$, $B^3_1 = <5>$, $B^2_2 = <6,5,6>$, $B^1_3 = <4,3,6,5,6,3,4>$, $B^4_1 = <7>$, $B^3_2 = <10,7,10>$, $B^2_3 = <12,5,10,7,10,5,12>$, $B^1_4 = <8,3,6,5,6,3,12,5,10,7,10,5,12,3,6,5,6,3,8>$.

We claim the following is true:

2). $\forall n,m \in \mathbf{N}$, $\text{rng } B^n_m \subseteq \{2^{m-1}(2k-1) : n \leq k \leq n+m\}$.

Certainly (2) is true if m = 1. Fix $n \in \mathbf{N}$ and suppose (2) is true $\forall m,i$, $m \in \mathbf{N}$, $i < m$. If $z < m$, then 2.3(2) shows $\alpha z \leq \log_2 m < m$. Hence, by induction,

$$\text{rng } B^{n+1}_{\alpha z} \subseteq \cup\{2^{m-1}(2k-1) : n+1 \leq k \leq n+m\}.$$

Because $2^{m-1}(2n-1)$ is the only element of $\text{rng } B^n_m$ not considered by the induction hypothesis, $\text{rng } B^n_m \subseteq \{2^{m-1}(2k-1) : n \leq k \leq n+m\}$.

We claim the following is true:

3). $\forall n \geq 0$, $\forall m,z \in \mathbf{N}$, $m \neq z$, no head or tail of B^n_m is a head or tail of B^n_z.

Suppose $n,m,z \in \mathbf{N}$, $z < m$. If $k \neq z$, then $2^{z-1}(2k-1) \notin I_k$. Thus, (2) shows, $2^{z-1}(2n-1) \notin \text{rng } B^n_m$. As $<2^{z-1}(2n-1)>$ is both a head and a tail of B^n_z, (3) is true.

$\forall n \in \mathbf{N}$, define $x_n = \alpha(B^n_z : z \in \mathbf{N})$. From 5.2, x_n is recurrent in $[\Pi;\sigma]$. Since $\forall m \in \mathbf{N}$, $m \leq 2^{m-1}(2n-1) = B^n_m(1)$, 4.7 proves:

4). $\forall n \in \mathbf{N}$, $\mathbf{OC}(x_n)$ is homeomorphic to $\mathbf{P}$.

$\forall n,p \in \mathbf{N}$, define

$$G_p = \{y \in \Pi : y|(1,p) = \alpha(B^{n+1}_z : z < 2^{p-1})\}.$$

Then $\{G_p : p \in \mathbf{N}\}$ forms a nhbd base at x_{n+1}. But (1) shows $\forall p \in \mathbf{N}$, $\exists m \in \mathbf{N}$ with $\alpha(B^{n+1}_z : z < p) \leq B^n_m = x_n|I$ for some interval I of $\mathbf{N}$. Hence, $\exists\, q \in \mathbf{N}$ such that $\sigma^q x_n \in G_p \cap \mathbf{Orb}(x_n)$. Thus, $x_{n+1} \in \mathbf{OC}(x_n)$. Similarly, $\forall r \in \mathbf{N}$, $\sigma^r x_{n+1} \in \mathbf{OC}(x_n)$. Further, since $x_n(1) = 2n-1 \notin \{2^{m-1}(2k-1) : k > n\}$, (2) shows $x_n \notin \mathbf{OC}(x_{n+1})$. So we have:

5). $\forall n \in \mathbf{N}$, $\mathbf{OC}(x_{n+1}) \subsetneq \mathbf{OC}(x_n)$.

Since $\forall n > 1$, $n < 2n-1$. (2) implies that $\forall n > 1$, $n < \min \text{rng } x_n$. Hence, the following holds:

6). $\forall k \in \mathbf{N}$, $\cap_{n > k} \mathbf{OC}(x_n) = \varnothing$.

Now fix $n \in \mathbf{N}$ and suppose that $y \in \mathbf{OC}(x_n)$, and either y is not recurrent or $\mathbf{OC}(y) \neq \mathbf{OC}(x_n)$. Using (3), 5.5 finds an $h \in \Pi$ such that $\forall z \in \mathbf{N}$, $y|[1,z] \leq B^n_{hz}$. So $\forall m \in \mathbf{N}$, $y|[1,m+1] \leq B^n_{h(\alpha(m))}$. From (1), $y|[1,z] \leq \alpha(B^{n+1}_{h(\alpha(m))} : 1 \leq z < h(\alpha(m)))$. Hence, $y \in \mathbf{OC}(x_{n+1})$. Applying (6) yields:

7). $\forall y \in \mathbf{OC}(x_0)$, y is stable, and $\exists n \geq 0$ such that $\mathbf{OC}(y) = \mathbf{OC}(x_n)$.

Finally notice that (5), (7), and 6.1 prove $\mathbf{OC}(x_0)$ contains no minimal sets. As (4) shows $\mathbf{OC}(x_0)$ is homeomorphic to $\mathbf{P}$, $\mathbf{OC}(x_0)$ satisfies our requirements. //

7. Multiple Recurrence.

In this section we are concerned with multiple systems - systems comprising of more than one map from a space to itself. One of the earliest results on multiple systems is due to P. Erdös and A. Stone [8]:

7.1. x *is recurrent (almost-periodic) in a system* [X;f] *iff* x *is jointly recurrent (almost-periodic) in the system* $[X;\{f^n : n\in \mathbf{N}\}]$.

More recent is the Furstenberg-Weiss theorem [9] (improved in [4]):

7.2. *If* $\mathfrak{F}$ *is a finite family of commuting maps on a compact metric space* X, *then* $[X;\mathfrak{F}]$ *has a multiply recurrent point.*

7.3. Lemma. *The following are true:*

1). x *is multiply recurrent in* $[\Pi,\{\sigma^1, \dots ,\sigma^n\}]$ *iff* $\forall m\in \mathbf{N}$, $\exists k = k_m\in \mathbf{N}$ *such that* $\forall p \le n$, $x|[kp+1,kp+m] \equiv x|[1,m]$.
2). x *is multiply recurrent in* $[\Pi,\{\sigma^n : n\in \mathbf{N}\}]$ iff $\forall m\in \mathbf{N}$, $\exists\, k = k_m\in \mathbf{N}$ *such that* $\forall n\in \mathbf{N}$, $x|[kn+1,kn+m] \equiv x|[1,m]$.
3). If x is multiply recurrent in $[\Pi,\{\sigma^n : n\in \mathbf{N}\}]$, then x is almost-periodic in $[\Pi;\sigma]$.

\\ (1) and (2) are immediate from the definition. Notice that (2) implies that $\forall m\in \mathbf{N}$, we can choose $k > m$ in the conclusion; hence, $\forall n\in \mathbf{N}$, $x|[1,m] \le x|[nk+1,(n+1)k]$. So (3) holds. //

7.4. Theorem. *Suppose* $(B_z : z\in \mathbf{N})$ *is a street and* $<\lambda B_z : z\in \mathbf{N}>$ *is a constant sequence. If* ϕ *is multiply recurrent in* $[\Pi,\{\sigma^1, \dots ,\sigma^n\}]$ *(in* $[\Pi,\{\sigma^n : n\in \mathbf{N}\}]$*),* then $\phi(B_z : z\in \mathbf{N})$ *is multiply recurrent in* $[\Pi,\{\sigma^1, \dots ,\sigma^n\}]$ (in $[\Pi,\{\sigma^n : n\in \mathbf{N}\}]$).

\\ The proof is analogous to the proof of 5.3(1). //

7.5. Corollary. *Suppose* $(B_z : z\in \mathbf{N})$ *is a street with* $<\lambda B_z : z\in \mathbf{N}>$ *a constant sequence. Then* $\alpha(B_z : z\in \mathbf{N})$ *is multiply recurrent in* $[\Pi,\{\sigma^n : n\in \mathbf{N}\}]$.

\\ $\forall m\in \mathbf{N}$, $\alpha|[1,2^{m-1}-1] = \alpha|[n\cdot 2^m+1,n\cdot 2^m+2^{m-1}-1] = (\sigma^n)^{2^m}\alpha|[1,2^{m-1}-1]$. So α is multiply recurrent in $[\Pi,\{\sigma^n : n\in \mathbf{N}\}]$. Now use 7.2. //

To see that 7.5 cannot be reversed try (1) of the next exercise. (2) shows that almost-periodic points are not necessary to get points multiply recurrent in each $[\Pi,\{\sigma^1, \dots ,\sigma^n\}]$.

7.6. EXERCISE.

1). *Prove* β, *the point defined in 5.4, is multiply recurrent in* $[\Pi,\{\sigma^n : n\in \mathbf{N}\}]$.

2). *Prove that the point* x *defined in 6.3 is,* $\forall n\in \mathbf{N}$, *a multiply recurrent point in* $[\mathbf{P},\{f^1, \dots ,f^n\}]$.

3). *Suppose* x *is the point defined in 6.3. Are all of the points in* $\mathbf{OC}(x)$ *multiply recurrent?*

7.7. Example. *There is a continuous function* $f: \mathbf{P} \to \mathbf{P}$ *such that* $[\mathbf{P},f]$ *is minimal and* $\forall y\in \mathbf{P}$, y *is not multiply recurrent in* $[\mathbf{P};\{f,f^2\}]$.

\\\\. Let $B_1 = <1>$. Define $x = \alpha(B_z : z\in \mathbf{N})$, where B_z is the constant function z of recursively defined length $\sum_{k=1}^{z-1} 2^{z-k}\lambda B_k$. The first "few" terms of x are $<1,2,2,1,3,3,3,3,1,2,2,1,4,4,4,4,4,4,4,4,4,4,4,4,1,2,2,1,3,3,3,3,1,2,2,1, \dots>$. Notice that $\lambda B_{\alpha(n)} = \lambda\alpha(B_z : z < 2^n)$ (see exercise 4.3).

From 3.2 $\mathbf{OC}(x)$ is homeomorphic to $\mathbf{P}$. 5.6 shows $[\mathbf{OC}(x);\sigma]$, and hence $[\mathbf{OC}(x);\{\sigma,\sigma^2\}]$, is minimal.

Suppose $y\in \mathbf{OC}(x)$. According to 5.6, y can be written as

$$((T\wedge\alpha(B_z : z < h(1)))\wedge\wedge_{n>1} (B_{\alpha(h(n))}\wedge\alpha(B_z : z\in [1,h(n))))),$$

where $h\in \Pi$ is increasing and T is a maximal tail of $B_{\alpha(h(1))}$ in y. Let $t = \lambda T$ and suppose $p\in \mathbf{N}$ satisfies $T \equiv y|[p+1,p+t]$. Since the B_z's have disjoint ranges,

$$p > \lambda(T\wedge\alpha(B_z : z < h(1))\wedge B_{\alpha(h(2))}).$$

Suppose $m\in \mathbf{N}$ is the largest integer such that $p \geq \lambda H$, where H is the block

$$((T\wedge\alpha(B_z : z < h(1)))\wedge(\wedge_{1<n<m} (B_{\alpha(h(n))}\wedge\alpha(B_z : z\in [1,h(n)))))\wedge B_{\alpha(h(m))}.$$

Then $\lambda B_{\alpha(h(m))} < p+1$. Since T and $B_{\alpha(h(m)+1)}$ have disjoint ranges,

$$p+1 < \sum_{k=1}^{m} 2^{m-k}\lambda B_{\alpha(n)} = 2\cdot\lambda B_{\alpha(h(m))}.$$

Thus, $\lambda(H)+\lambda B_{\alpha(h(m))} < 2p+1 \leq 3\cdot\lambda B_{\alpha(h(m))} = \lambda B_{\alpha(h(m)+1)} \leq \lambda B_{\alpha(h(m+1))}$. Thus, $x(2p+1)\in \operatorname{rng} B_{\alpha(h(m+1))}$. Since T and $B_{\alpha(h(m)+1)}$ have disjoint ranges,

$$\sigma(y(p)) = x(p+1) \neq x(2p+1) = \sigma^2(y(p)).$$

Therefore, y is not multiply recurrent in $[\mathbf{OC}(x);\{\sigma,\sigma^2\}]$. //

7.8. Conjectures.

1. *Suppose* x *is an almost-periodic point in* $[\Pi,\sigma]$ *and suppose* $\mathfrak{F} \subseteq \{\sigma^n : n\in \mathbf{N}\}$ *is finite. Then* $[\mathbf{OC}(x),\mathfrak{F}]$ *has a multiply recurrent point.*
2. *There is a completely metrizable space* X *and an almost-periodic point* x *in the system* $[X,f]$ *such that each point of* X *fails to be multiply recurrent in* $[X;\mathfrak{F}]$ *for some finite* $\mathfrak{F} \subseteq \{f^n : n\in \mathbf{N}\}$.

7.9. Example. *There are commuting homeomorphisms* $f,g : \mathbb{P} \to \mathbb{P}$ *such that* $[\mathbf{P},\{f,g\}]$ *is minimal,* $\forall p\in \mathbf{P}$, $\mathbf{OC}_f(p)\cap\mathbf{OC}_g(p) = \{p\}$.

\\\\ In Π, let $P = \mathbf{OC}(\alpha)$. 4.4 shows P is homeomorphic to $\mathbf{P}$. From 3.2, P^2 is homeomorphic to $\mathbf{P}$. For the identity map $f\in\Pi$, let $f = \sigma\times id$ and $g = id\times\sigma$. Then f and g commute. From 2.3(a) and 6.2, $[P;\sigma]$ is minimal; hence, $[P^2;\{f,g\}]$ is a minimal multiple system. Clearly, $\forall (x,y)\in P^2$, $\mathbf{OC}_f((x,y)) = P\times\{y\}$ and $\mathbf{OC}_g((x,y)) = \{x\}\times P$. So $\mathbf{OC}_f((x,y))\cap\mathbf{OC}_g((x,y)) = \{(x,y)\}$. //

REFERENCES:

[1] E. Akin, *The General Topology of Dynamical Systems,* Graduate Studies in Mathematics 1, American Mathematical Society, 1993.

[2] E. Akin, J. Auslander, and K. Berg, *When in a transitive map chaotic?,* to appear.

[4] B. Balcar, P. Kalasek, and S. Williams, *Multiple Recurrence in dynamical systems,* Comment. Math. Univ. Carolina **28** (1987), 607-612.

[6] R. Ellis, *Lectures on Topological Dynamics,* W.A. Benjamin, Inc. (1969).

[7] R. Engelking, *General Topology,* Polish Scientific Publishers (1977).

[8] P. Erdös and A.H. Stone, *Some remarks on almost-periodic transformations* , Bulletin AMS **51** (1945), 126-130

[9] H. Furstenberg, *Recurrence in Ergodic Theory and Combinatorial Number Theory,* Princeton Univ. Press (1981).

[10] S. Glasner and D. Maon, *Rigidity in topological dynamics,* Ergodic Theory and Dynamical Systems 9 (1989), 177-188.

[11] W. Gottschalk, *Orbit-closure decompositions and almost periodic properties* , Bull. AMS **50** (1944), 915-919.

[12] W. Gottschalk and G. Hedlund, Topological Dynamics, Amer. Math. Soc. Colloquium Pub. **36** (1955).

[13] G. Hedlund, *Transformations commuting with the shift* , Topological Dynamics, W. A. Benjamin (1966), 259-290.

[14] G. Hedlund, *Endomorphisms and automorphisms of the shift dynamical system,* Math. Systems Theory **3** (1969), 320-375.

[15] Y. Katznelson and B. Weiss, *When all points are recurrent/generic,* Ergodic Theory and Dynamical Systems I Proceedings, Special Year Maryland 1979-80.

[16] J. Pelant and S. Williams, *Examples on recurrence,* Annals of the New York Academy of Sciences 806 (1996), 316-332.

[17] K. Petersen, *Ergodic Theory,* Cambridge University Press (1983).

[18] S. Williams, *Special points arising from self-maps,* General Topology and relations to Modern Analysis **5** (1988), 629-638.

State University of New York at Buffalo, Buffalo, N.Y. 14214 U.S.A.
E-mail addresses: sww@acsu.buffalo.edu or BONVIBRE@AOL.COM

Part II

Poster Presentations

DIMACS Series in Discrete Mathematics
and Theoretical Computer Science
Volume **34**, 1997

Finding elliptic curves defined over Q of high rank

Garikai Campbell

ABSTRACT. We describe the best known method for constructing curves defined over the rationals whose Mordell-Weil group has high rank. The idea, due to Mestre, is to find an elliptic curve defined over the rational function field $Q(t)$ of rank ≥ 11 and then to specialize at a particular rational value of t to produce an elliptic curve defined over Q. By choosing t appropriately, the curve will have significantly higher rank.

1. General construction

Mestre's method for producing elliptic curves of rank at least 11 over $Q(t)$ can be summarized in the following two propositions.

PROPOSTION 1.1. ([1]) *If* $\mathbf{q}(\mathbf{x}) = \prod_{\mathbf{i}=\mathbf{1}}^{\mathbf{6}}(\mathbf{x} - \mathbf{a_i})$ *with* $\mathbf{a_i} \in \mathbf{Z}$ *and* $\mathbf{p}(\mathbf{x}) = \mathbf{q}(\mathbf{x} - \mathbf{t})\mathbf{q}(\mathbf{x} + \mathbf{t})$ *then we can write* $\mathbf{p}(\mathbf{x}) = \mathbf{g}(\mathbf{x})^{\mathbf{2}} - \mathbf{r}(\mathbf{x})$ *where*

1. $\mathbf{g}, \mathbf{r} \in \mathbf{Q}[\mathbf{x}]$, *and*
2. $\mathbf{g}$ *is of degree 6, and* $\mathbf{r}$ *is of degree* ≤ 5.

PROPOSTION 1.2. ([1]) *If in the above, the* $\mathbf{a_i}$ *are chosen suitably,* $\mathbf{r}$ *is of degree 4 and* $\mathbf{y^2} = \mathbf{r}(\mathbf{x})$ *is an elliptic curve containing the twelve points* $(\mathbf{a_i} + \mathbf{t}, \mathbf{g}(\mathbf{a_i} + \mathbf{t}))$ *and* $(\mathbf{a_i} - \mathbf{t}, \mathbf{g}(\mathbf{a_i} - \mathbf{t})), \mathbf{i} = \mathbf{1}, \mathbf{2}, \ldots, \mathbf{6}$.

This particular construction does not provide a way of choosing the a_i so that $r(x)$ is of degree 4, nor does it guarantee that once suitable a_i have been found, the twelve points of proposition 1.2 will be independant over $Q(t)$. Mestre does prove that an analagous construction yields a family of elliptic curves over $Q(t)$ for which these twelve points are independant. In spite of this, we chose to use the former construction for two reasons.

First, although the twelve points produced in this construction need not be independant, in practice, they almost always are. Second, this construction simplifies many of the computations. In Mestre's other construction, the resulting curve is in general a cubic in x and y with both the x^3 and y^3 terms appearing. We can find rational points and compute $a_p = p + 1 - \#E(F_p)$, two neccessary tasks, more efficiently on curves of the form $y^2 = r(x)$. It is important to note that in each

1991 *Mathematics Subject Classification.* 11G05.

construction, the size of the coefficients of the curve in Weierstrass form prohibit efficient calculations. Therefore, we perform nearly all computations on the curve in its initial form of $y^2 = r(x)$, where $r(x)$ is a quartic.

If the twelve points in this construction are linearly independant, then by taking one to be the identity of the group, we have an elliptic curve of rank at least 11 over $Q(t)$. Mestre [2] and Nagao [3] have since extended this result and found elliptic curves over $Q(t)$ of rank ≥ 12 and ≥ 13, respectively.

2. Finding specific curves

Once an elliptic curve, E_t, defined over $Q(t)$ has been constructed, the next step is to find a rational number t_0 so that E_{t_0} is an elliptic curve defined over Q with even higher rank. Mestre [4], Fermigier [5] and Nagao [6] have illustrated that experimentally the sum

$$S(E,N) = \sum_{\substack{p \leq N \\ p\ prime}} \frac{-(2+a_p)}{p+1-a_p} \log p$$

is large when the rank of E is large. Conversely, we have found that if $S(E, p_{90}) \geq 38$, where p_{90} is the 90th prime, it is a good indication that E has large rank. We hope to use this information to find an elliptic curve over Q of rank greater than 21. Nagao and Kouya found an elliptic curve of rank at least 21, the largest known rank [7].

3. Current Approach

We currently search for such a curve as follows. We first construct an elliptic curve, E_t, over $Q(t)$ containing twelve points. We then compute $S(E_{t_0}, p_{90})$ for all t_0 in some subset of the rationals. This subset is determined by what is computationally feasible given the coefficeints of the curve E_t. We search for points on the curves for which this sum is greater than 38 and then compute which of those points are linearly independant. This final step is accomplished by computing the determinant of the matrix $(< P_i, P_j >)_{1 \leq i,j \leq n}$, where $<,>$ is the canonical height pairing. If this determinant is nonzero, the points $\{P_i\}_{i=1}^n$ are linearly independant. We hope to expand on this idea by eliminating the step of plugging in rational values of t and working completely over the function field. In so doing, we are attempting to find a curve of larger rank over $Q(t)$. This will give a better starting point for finding an elliptic curve defined over Q of rank at least 22.

4. References

1 J.F. Mestre, *Courbes elliptiques de rang $\geq$ 11 sur $Q(t)$*. C.R. Acaad. Sci. Paris **313**, Ser. 1 (1991), 139-142.

2 J.F. Mestre, *Courbes elliptiques de rang $\geq$ 12 sur $Q(t)$*. C.R. Acad. Sci. Paris **313**, Ser. 1 (1991), 171-174.

3 K. Nagao, *An Example of Elliptic Curve over $Q(t)$ with rank* ≥ 13. Proc. Japan Acad. **70**, Ser. A (1994), 152-153.

4 J.F. Mestre, *Construction de coubes elliptiques sur Q de rang* ≥ 12. C.R. Acad. Sci. Paris **295**, Ser. 1 (1982), 643-644.

5 S. Fermigier, *Un example de courbe elliptique definie sur Q de rang* ≥ 19. C.R. Acad. Sci. Paris **t. 315**, Ser. 1 (1992), 719-722.

6 K. Nagao, *An Example of Elliptic Curv. over Q with rank* $\geq$ 20. Proc. Japan Acad. **70**, Ser. A (1993), 291-293.
7 K. Nagao and T. Kouya, *An Example of Elliptic Curve over Q with rank* $\geq$ 21. Proc. Japan Acad. **70**, Ser. A (1994), 104-105.

DEPARTMENT OF MATHEMATICS, RUTGERS UNIVERSITY, NEW BRUNSWICK, NJ 08901
Current address: Department of Mathematics, Rutgers University, New Brunswick, NJ 08901
E-mail address: gcampbel@math.rutgers.edu

DIMACS Series in Discrete Mathematics
and Theoretical Computer Science
Volume **34**, 1997

SYMPLECTIC MATRIX STRUCTURE IN NUMERICAL INTEGRATION

Michael Keeve

ABSTRACT. In recent years, a lot of work has been done on symplectic integration. One issue not addressed in these works is the one of maintaining symplecticity in finite precision (i.e., at the linear algebra level). This paper is a step in this direction. Applications from control theory, in particular the integration of the Differential Riccati Equation (DRE) motivated our research.

1. Introduction

In this paper, we consider the following problem: matrices $D, N \in \Re^{2N\times 2N}$ are given such that $D^{-1}N$ is symplectic, and we devise a way to solve $DX = N$ so that symplecticity is maintained in spite of numerical error.

Consider a *s stage Gauss Runge-Kutta (GRK)* scheme for linear constant coefficient Hamiltonian systems. That is, we have the real $2n$-dimensional linear system (the superscript "t" denotes transposition)

$$\dot{\mathbf{y}} = H\mathbf{y}, \quad \text{where} \quad (JH)^t = JH, \; J = \begin{pmatrix} 0 & I \\ -I & 0 \end{pmatrix}, \quad \text{or} \quad H = \begin{pmatrix} A & C \\ B & -A^t \end{pmatrix} \quad B = B^t, \; C = C^t. \tag{1.1}$$

Clearly, the exact solution is $\mathbf{y}(t) = e^{tH}\mathbf{y}_0$, and it is well understood that e^{tH} is a symplectic matrix for all times t. Recall that a matrix S is symplectic if $S^tJS = J$, or -equivalently- if $S^{-1} = -JS^tJ$. Given $\mathbf{y}_0$ and h, we seek $\mathbf{y}_1 = \mathbf{y}_0 + h\sum_{i=1}^{s} b_i\mathbf{k}_i$ where $\mathbf{k}_i = H\left(\mathbf{y}_0 + h\sum_{j=1}^{s} a_{ij}\mathbf{k}_j\right)$, $i = 1, 2, \ldots, s$. The RK coefficients b_i, a_{ij} are the usual ones of Gauss schemes, in particular $\sum_{j=1}^{s} a_{ij} = c_i$, the i-th zero of the Legendre polynomial of degree s renormalized on $[0,1]$, and $b_i > 0$. It is well known that, in exact arithmetic, $\mathbf{y}_1 = \Phi_s(hH)\mathbf{y}_0$, $\Phi_s(hH)$ is the (s,s) diagonal Padè approximant to e^{hH}, and $\Phi_s(hH)$ is symplectic. In what follows, we will assume that h is such that $\Phi_s(hH)$ is well defined, will absorb the dependence on h into H, and will write

$$\Phi_s(H) = (D_s(H))^{-1}N_s(H) \tag{1.2}$$

where D_s and N_s are the "denominator" and "numerator" of the Padè approximation. Thus, the need arises to solve a linear system in order to find $\Phi_s(H)$. To this end, the default choice is to use a Gaussian elimination based approach. This is a non-symplectic linear algebra procedure, and as a consequence symplecticity is usually lost. In our experience, this loss is directly proportional to the norm of the inverse of the system to be solved, that is to $\|(D_s(H))^{-1}\|$. Thus, this is no major concern if h is small, since in such a case

Work supported by NSF Grants DMS-9306412, DMS-9625813. 1991 *Mathematics Subject Classification.* Primary 65F99; Secondary 65L99.

$D_s(H) \approx I$, but it may be a concern for large h. We typically need to take large h when the above setup is used in conjunction with the solution process for the Differential Riccati Equation(DRE).

$$\dot{X} = AX + XA^t + C - XBX\,,$$

via the relation

$$X = YZ^{-1}\,, \qquad \begin{pmatrix} \dot{Y} \\ \dot{Z} \end{pmatrix} = H \begin{pmatrix} Y \\ Z \end{pmatrix}.$$

(Large h arises because we wish to control errors in X, but not necessarily in Y, Z.) We refer to [DE1-2] for justification of the above setup; here, it suffices to remark that the results in [DE1-2] necessitates that the discrete transition matrix carrying out the approximation for $\begin{pmatrix} Y \\ Z \end{pmatrix}$ be symplectic.

In this note, our point of view will be to obtain directly an expression for $\Phi_s(H)$ as product of elementary symplectic matrices, for which finite precision symplecticity is more or less trivially enforced. Although we make no claims as to the (backward) numerical stability of the resulting procedure, our experience on problems originating from DREs indicates that they provide accurate, as well as symplectic, approximations.

2. Factorization of Transition Matrix

We will make the following **simplifying assumptions**

(A0) Without loss of generality (as it will be clear from the proofs below), let us assume that the stage value s is an odd number: $s = 2k+1$, $k = 0, 1, \ldots$.

(A1) The Hamiltonian matrix H is of the following form

$$H := \begin{pmatrix} H_{11} & H_{12} \\ H_{21} & H_{22} \end{pmatrix} = \begin{pmatrix} A & T \\ D & -A \end{pmatrix}, \tag{2.1}$$

where A and D are diagonal and T is symmetric tridiagonal. We will henceforth assume that D is invertible. This form is then known as "unreduced J-tridiagonal form"; the J-tridiagonal form can be obtained (essentially all the time) via symplectic transformations (see [BMW] for details of this). So, we can think of having H as in (2.1) via a preprocessing step.

There are a number of properties of the powers of H, H^{2k} and H^{2k+1}, which we will repeatedly use. They are not hard to verify, and can be found in [K]. With obvious block notation, we will use the following:

$$H^{2k} = \begin{pmatrix} (H^{2k})_{11} & (H^{2k})_{12} \\ 0 & (H^{2k})^t_{11} \end{pmatrix}, \qquad (H^{2k})^t_{12} = -(H^{2k})_{12}\,, \qquad k = 0, 1, \ldots, \tag{2.2}$$

$$H^{2k+1} = \begin{pmatrix} (H^{2k+1})_{11} & (H^{2k+1})_{12} \\ (H^{2k+1})_{21} & -(H^{2k+1})^t_{11} \end{pmatrix}, \qquad (JH^{2k+1})^t = JH^{2k+1}\,, \qquad k = 0, 1, \ldots. \tag{2.3}$$

Now, (2.3) states that H^{2k+1} is Hamiltonian, and therefore $(H^{2k+1})_{12}$ and $(H^{2k+1})_{21}$ must be symmetric matrices. Since

$$H^{2k+1} = (H)(H^{2k}) = \begin{pmatrix} H_{11}(H^{2k})_{11} & H_{11}(H^{2k})_{12} + H_{12}(H^{2k})^t_{11} \\ H_{21}(H^{2k})_{11} & H_{21}(H^{2k})_{12} - H^t_{11}(H^{2k})^t_{11} \end{pmatrix}$$

we then have

$$H_{21}(H^{2k})_{11} = (H^{2k})^t_{11}H_{21}\,, \qquad (H^{2k})_{11}(H_{21})^{-1} = (H_{21})^{-1}(H^{2k})^t_{11}\,, \quad k = 0,1,\dots. \tag{2.4}$$

From the form of the even powers of H in (2.2), we also have

$$\begin{aligned}(H^{2k+2})_{11} &= (H^{2k})_{11}(H^2)_{11} = \dots = \overbrace{(H^2)_{11}\dots(H^2)_{11}}^{(k+1)\text{ times}} \qquad \text{and}\\ (H^{2k+2})_{12} &= (H^{2k})_{11}(H^2)_{12} + (H^{2k})_{12}(H^2)^t_{11}\,, \quad k = 0,1,\dots.\end{aligned} \tag{2.5}$$

For $\Phi_s(H)$ in (1.2), it is convenient to rewrite the "numerator and denominator" as follows

$$N_{2k+1}(H) = \begin{pmatrix} Y & Q \\ F & -X \end{pmatrix}, \qquad D_{2k+1}(H) = \begin{pmatrix} V & R \\ -F & -W \end{pmatrix}, \tag{2.6}$$

$$\begin{aligned} F &= d_1H_{21} + d_3(H^3)_{21} + \dots + d_{2k+1}(H^{2k+1})_{21}\,,\\ -X &= I - d_1H^t_{11} + d_2(H^2)^t_{11} - \dots + d_{2k}(H^{2k})^t_{11} - d_{2k+1}(H^{2k+1})^t_{11}\,,\\ Y &= I + d_1H_{11} + d_2(H^2)_{11} + \dots + d_{2k}(H^{2k})_{11} + d_{2k+1}(H^{2k+1})_{11}\,,\\ Q &= d_1H_{12} + d_2(H^2)_{12} + \dots + d_{2k}(H^{2k})_{12} + d_{2k+1}(H^{2k+1})_{12}\,,\\ -W &= I + d_1H^t_{11} + d_2(H^2)^t_{11} + \dots + d_{2k}(H^{2k})^t_{11} + d_{2k+1}(H^{2k+1})^t_{11}\,,\\ V &= I - d_1H_{11} + d_2(H^2)_{11} - \dots + d_{2k}(H^{2k})_{11} - d_{2k+1}(H^{2k+1})_{11}\,,\\ R &= -d_1H_{12} + d_2(H^2)_{12} - \dots + d_{2k}(H^{2k})_{12} - d_{2k+1}(H^{2k+1})_{12}\,.\end{aligned} \tag{2.7}$$

We will now make the following **structural assumption** :

(A2) $F = d_1H_{21} + d_3(H^3)_{21} + \dots + d_{2k+1}(H^{2k+1})_{21}$ be invertible for $k = 0,1,\dots$.

We now can formulate

THEOREM 2.1. (Factorization of $\Phi(H)$). *Under assumptions (A0)-(A2), the matrix $\Phi_s(H)$ admits the following factorization*

$$\Phi_s(H) = -\begin{pmatrix} I & -F^{-1}W \\ 0 & I \end{pmatrix}\begin{pmatrix} I & 0 \\ -K^{-1}(V+Y) & I \end{pmatrix}\begin{pmatrix} I & -F^{-1}X \\ 0 & I \end{pmatrix}, \tag{2.8}$$

where

$$K = R - VF^{-1}W\,. \tag{2.9}$$

REMARK 2.2. Before verifying (2.8), notice that the matrices appearing there are of the form $\begin{pmatrix} I & E \\ 0 & I \end{pmatrix}$, or $\begin{pmatrix} I & 0 \\ G & I \end{pmatrix}$, which are clearly symplectic, if $E = E^t$ and $G = G^t$. Because of Lemma 2.3 below, this is the case for us. Therefore, in practice, to maintain symplecticity in finite precision, it will suffice to guarantee that the matrices $-F^{-1}W$, $-K^{-1}(V+Y)$, and $-F^{-1}X$, stay symmetric, which is not hard to enforce.

Proof. With previous notation, we have

$$D_s(H) = \begin{pmatrix} V & R \\ -F & -W \end{pmatrix} = -JJ\begin{pmatrix} V & R \\ -F & -W \end{pmatrix},$$

and thus

$$D_s^{-1}(H) = \left[-JJ\begin{pmatrix} V & R \\ -F & -W \end{pmatrix}\right]^{-1} = \left[J\begin{pmatrix} F & W \\ V & R \end{pmatrix}\right]^{-1} = \begin{pmatrix} F & W \\ V & R \end{pmatrix}^{-1}(-J)$$

Since F is invertible, $\begin{pmatrix} F & W \\ V & R \end{pmatrix}^{-1}$ can be written as $\begin{pmatrix} I & -F^{-1}WK \\ 0 & K^{-1} \end{pmatrix}\begin{pmatrix} F^{-1} & 0 \\ -VF^{-1} & I \end{pmatrix}$ where K is defined by (2.9) and is invertible (Schur complement of invertible matrix). Similarly

$$-JN_s(H) = -J\begin{pmatrix} Y & Q \\ F & -X \end{pmatrix} = \begin{pmatrix} -F & X \\ Y & Q \end{pmatrix} = \begin{pmatrix} -F & 0 \\ Y & I \end{pmatrix}\begin{pmatrix} I & -F^{-1}X \\ 0 & Q+YF^{-1}X \end{pmatrix}.$$

Therefore,

$$\begin{aligned}
\Phi_s = D_s^{-1}N_s &= \begin{pmatrix} I & -F^{-1}WK \\ 0 & K^{-1} \end{pmatrix}\begin{pmatrix} F^{-1} & 0 \\ -VF^{-1} & I \end{pmatrix}\begin{pmatrix} -F & 0 \\ Y & I \end{pmatrix}\begin{pmatrix} I & -F^{-1}X \\ 0 & Q+YF^{-1}X \end{pmatrix} \\
&= \begin{pmatrix} I & -F^{-1}W \\ 0 & I \end{pmatrix}\begin{pmatrix} I & 0 \\ 0 & K^{-1} \end{pmatrix}\begin{pmatrix} -I & 0 \\ V+Y & I \end{pmatrix}\begin{pmatrix} I & 0 \\ 0 & Q+YF^{-1}X \end{pmatrix}\begin{pmatrix} I & -F^{-1}X \\ 0 & I \end{pmatrix} \\
&= \begin{pmatrix} I & -F^{-1}W \\ 0 & I \end{pmatrix}\begin{pmatrix} -I & 0 \\ K^{-1}(V+Y) & K^{-1} \end{pmatrix}\begin{pmatrix} I & 0 \\ 0 & -K \end{pmatrix}\begin{pmatrix} I & -F^{-1}X \\ 0 & I \end{pmatrix} \\
&= -\begin{pmatrix} I & -F^{-1}W \\ 0 & I \end{pmatrix}\begin{pmatrix} I & 0 \\ -K^{-1}(V+Y) & I \end{pmatrix}\begin{pmatrix} I & -F^{-1}X \\ 0 & I \end{pmatrix},
\end{aligned}$$

where we have used $Q + YF^{-1}X = -K$ from (iii) of Lemma 2.3 below. □

To complete the proof of Theorem 2.1, we need to show the following Lemma.

LEMMA 2.3. *Under assumptions (A0)-(A2), the following properties hold*

(i) $-F^{-1}X$ *is symmetric,*

(ii) $-F^{-1}W$ *is symmetric,*

(iii) $Q + YF^{-1}X = -K$,

(iv) $-K^{-1}(V+Y)$ *is symmetric or* $\begin{pmatrix} I & 0 \\ -K^{-1}(V+Y) & I \end{pmatrix}$ *is symplectic.*

Proof. We begin showing (i). From the explicit form of $(H^{2k+1})_{11}$ and $(H^{2k+1})_{21}$, $k = 0, 1, \ldots$, (2.4), and elementary manipulation, we have

$$\begin{aligned}
-F^{-1}X = &-(H_{21})^{-1}H_{11}^t + (d_1 I + d_3(H^2)_{11} + \cdots + d_{2k+1}(H^{2k})_{11})^{-1} \\
&(I + d_2(H^2)_{11} + \cdots + d_{2k}(H^{2k})_{11})(H_{21})^{-1}.
\end{aligned} \tag{2.10}$$

Now, H_{21} and H_{11} are diagonal, and hence $(H_{21})^{-1}H_{11}^t$ is symmetric. To show that

$$(d_1 I + d_3(H^2)_{11} + \cdots + d_{2k+1}(H^{2k})_{11})^{-1}((H_{21})^{-1} + d_2(H^2)_{11}(H_{21})^{-1} + \ldots + d_{2k}(H^{2k})_{11}(H_{21})^{-1})$$

is symmetric, we show that

$$(d_1 I + d_3(H^2)_{11} + \cdots + d_{2k+1}(H^{2k})_{11})^{-1}(H^{2j})_{11}(H_{21})^{-1} \quad \text{is symmetric} \tag{2.11}$$

for any j, from which (i) will then follow. Now, taking the transpose of the expression in (2.11), and using (2.4) twice, we get

$$(H^{2j})_{11}\left[d_1 I + \cdots + d_{2k+1}(H^{2k})_{11}\right]^{-1} H_{21}^{-1},$$

and since trivially

$$(H^{2j})_{11}\left[d_1 I + \cdots + d_{2k+1}(H^{2k})_{11}\right]^{-1} = \left[d_1 I + \cdots + d_{2k+1}(H^{2k})_{11}\right]^{-1}(H^{2j})_{11},$$

we are done. To show (ii) is immediate, since (like in the previous argument for (i))

$$\begin{aligned} -F^{-1}W = &(H_{21})^{-1}H_{11}^t + (d_1 I + d_3(H^2)_{11} + \cdots + d_{2k+1}(H^{2k})_{11})^{-1} \\ &(I + d_2(H^2)_{11} + \cdots + d_{2k}(H^{2k})_{11})(H_{21})^{-1}, \end{aligned} \tag{2.12}$$

and therefore (ii) follows from (i). Next, we show (iii), or

$$Q + R + YF^{-1}X - VF^{-1}W = 0.$$

Using (2.9), (2.10), (2.12), and elementary manipulation, we have

$$\begin{aligned} &Q + R + YF^{-1}X - VF^{-1}W = 2d_2(H^2)_{12} + \cdots + 2d_{2k}(H^{2k})_{12} + 2(I + \cdots + d_{2k}(H^{2k})_{11})(H_{21})^{-1}H_{11}^t \\ &- (I + d_1 H_{11} + \cdots + d_{2k}(H^{2k})_{11} + d_{2k+1}H_{11}(H^{2k})_{11}) \\ &(d_1 I + d_3(H^2)_{11} + \cdots + d_{2k+1}(H^{2k})_{11})^{-1}(I + d_2(H^2)_{11} + \cdots + d_{2k}(H^{2k})_{11})(H_{21})^{-1} \\ &+ (I - d_1 H_{11} + d_2(H^2)_{11} - \cdots + d_{2k}(H^{2k})_{11} - d_{2k+1}H_{11}(H^{2k})_{11}) \\ &(d_1 I + d_3(H^2)_{11} + \cdots + d_{2k+1}(H^{2k})_{11})^{-1}(I + d_2(H^2)_{11} + \cdots + d_{2k}(H^{2k})_{11})(H_{21})^{-1}, \end{aligned}$$

which, using $(H^{2k+1})_{11} = H_{11}(H^{2k})_{11}$ and some algebra, further simplifies to

$$\begin{aligned} &Q + R + YF^{-1}X - VF^{-1}W = \\ &= 2d_2(H^2)_{12} + \cdots + 2d_{2k}(H^{2k})_{12} + 2(I + d_2(H^2)_{11} + \cdots + d_{2k}(H^{2k})_{11})(H_{21})^{-1}H_{11}^t \\ &- 2H_{11}(I + d_2(H^2)_{11} + \cdots + d_{2k}(H^{2k})_{11})(H_{21})^{-1} \\ &= 2d_2(H^2)_{12} + \cdots + 2d_{2k}(H^{2k})_{12} + 2(I + d_2(H^2)_{11} + \cdots + d_{2k}(H^{2k})_{11})(H_{21})^{-1}H_{11}^t \,. \\ &- 2H_{11}(H_{21})^{-1}(I + d_2(H^2)_{11}^t + \cdots + d_{2k}(H^{2k})_{11}^t) \\ &= 2d_2(H^2)_{12} + \cdots + 2d_{2k}(H^{2k})_{12} + 2(d_2(H^2)_{11} + \cdots + d_{2k}(H^{2k})_{11})(H_{21})^{-1}H_{11}^t \\ &- 2H_{11}(H_{21})^{-1}(d_2(H^2)_{11}^t + \cdots + d_{2k}(H^{2k})_{11}^t) \end{aligned}$$

We now show by induction that

$$(H^{2k+2})_{12} + (H^{2k+2})_{11}(H_{21})^{-1}H_{11}^t - H_{11}(H_{21})^{-1}(H^{2k+2})_{11}^t = 0 \quad k = 0, 1, \ldots . \tag{2.13}$$

from which the above expression for $Q + R + YF^{-1}X - VF^{-1}W$ will vanish identically, and (iii) will follow. Now, clearly (2.13) is true for $k = 0$, since in this case it is

$$\begin{aligned} &(H^2)_{12} + (H^2)_{11}(H_{21})^{-1}H_{11}^t - H_{11}(H_{21})^{-1}(H^2)_{11}^t \\ &= AT - TA + A^3D^{-1} + TA - A^3D^{-1} - AT = 0. \end{aligned} \tag{2.14}$$

We now assume (2.13) for $k = m - 1$, and show it for $k = m$. We have

$$
\begin{aligned}
&(H^{2m+2})_{12} + (H^{2m+2})_{11}(H_{21})^{-1}H_{11}^t - H_{11}(H_{21})^{-1}(H^{2m+2})_{11}^t = \\
&(H^{2m})_{11}(H^2)_{12} + (H^{2m})_{12}(H^2)_{11}^t + (H^{2m})_{11}(H^2)_{11}(H_{21})^{-1}H_{11}^t - H_{11}(H_{21})^{-1}((H^2)_{11}(H^{2m})_{11})^t \\
&= [(H^{2m})_{12} - H_{11}(H_{21})^{-1}(H^{2m})_{11}^t](H^2)_{11}^t + (H^{2m})_{11}[(H^2)_{12} + (H^2)_{11}(H_{21})^{-1}H_{11}^t] \\
&= -(H^{2m})_{11}(H_{21})^{-1}H_{11}^t(H^2)_{11}^t + (H^{2m})_{11}H_{11}(H_{21})^{-1}(H^2)_{11}^t = 0,
\end{aligned}
$$

where we have used (2.14) and the induction hypothesis. Finally, we need to show (iv). In such case, we can argue as follows. We know that Φ_s is symplectic and we have the explicit factorization (2.8) for it. Because of (i)-(ii) of this Lemma, also $\begin{pmatrix} I & -F^{-1}W \\ 0 & I \end{pmatrix}$ and $\begin{pmatrix} I & -F^{-1}X \\ 0 & I \end{pmatrix}$ are symplectic. Therefore,

$$
-\begin{pmatrix} I & F^{-1}W \\ 0 & I \end{pmatrix} \Phi_s \begin{pmatrix} I & F^{-1}X \\ 0 & I \end{pmatrix}
$$

must be symplectic because symplectic matrices are closed under matrix multiplication. As a consequence, $\begin{pmatrix} I & 0 \\ -K^{-1}(V+Y) & I \end{pmatrix}$ is symplectic and thus $-K^{-1}(V+Y)$ must be symmetric. □

REMARK 2.4. As a final remark, we point out that in the case of $s = 1$, i.e. the implicit midpoint rule, the factorization (2.8) is valid also in case the Hamiltonian matrix H is time dependent (since it only gets evaluated at the midpoint); for general s, and time dependent H, the issue of a full symplectic factorization of the Gauss RK transition matrix is instead still an open problem.

Acknowledgments. I would like to thank Prof. Luca Dieci for his guidance and invaluable contributions and Prof. Timo Eirola for many insightful discussions.

3. References

[BMW] Bunse-Gerstner, A., Mehrmann, V., Watkins, D.: "An SR Algorithm For Hamiltonian Matrices Based on Gaussian Elimination", *Methods of Operations Research* **58** (1989), pp. 339-358.

[DE1] Dieci, L. and Eirola, T.: "Positive Definiteness in the Numerical Solution of Riccati Differential Equations", *Numerische Mathematik* **67** (1994), pp. 303-313.

[DE2] Dieci, L. and Eirola, T.: "Preserving Monotonicity in the Numerical Solution of Riccati Differential Equations", *Numerische Mathematik* to appear (1996).

[HW] Hairer, E. and Wanner, G.: *Solving Ordinary Differential Equations II* , Springer-Verlag, New York (1991).

[K] Keeve, M.: "Gauss Runge-Kutta Schemes for Initial Value Problems of Ordinary Differential Equations", *Ph. D. Thesis* in progress.

Mathematics Department, Georgia Institute of Technology, Atlanta, Georgia, 30332-0160.

E-mail address: mkeeve@math.gatech.edu

DIMACS Series in Discrete Mathematics
and Theoretical Computer Science
Volume **34**, 1997

A Numerical Algorithm for the Computation of Invariant Circles

Kossi Edoh

ABSTRACT. In this work we present a numerical method to compute invariant circles for maps. It is an application of the Hadamard graph transform (HGT) method for computing attracting invariant manifolds. We perform a sequence of discrete graph transforms, whereby for Poincaré maps each transform requires the solution of finitely many boundary value problems. We test the algorithm on two problems and observe the breakdown of their invariant circles.

1. Introduction

The computation of invariant set has been a major concern in the area of dynamical systems. A set $M \subset \mathbb{R}^n$ is said to be invariant under the map P if for any $x_0 \in M$ we have $P^n(x_0) \in M$ for all n. A similar definition exists for flows (see [W] page 14). One important low dimensional manifold is the invariant circle for maps. An invariant torus for flows can be defined implicitly from an invariant circle for a Poincaré map. We introduce an algorithm to compute the invariant circle of discrete maps such as Poincaré maps. We extend the algorithm to the computation of the invariant two-torus for flows. This is useful during the breakdown of the torus. At this stage the torus naturally loses its smoothness and becomes harder to compute. By computing an invariant circle for the Poincaré map instead of the torus for the flows, one can thereby reduce the complexity of the torus computation.

In [ACHM], a direct iteration is used to compute the invariant circle of discrete maps. When the circle contains fixed points many iterations are needed to capture the whole circle. Some of the recent methods to compute invariant circles are the Poincaré map approaches of [VV, KASP] and the method of characteristics of Dieci et al [DL]. The method of van Veldhuizen is based on the interpolation of points on the circle. His algorithm suffers from the unbounded norm of higher order interpolation schemes on unequal mesh on the circle. The algorithm thus requires the use of a large number of mesh points. The algorithm of Kevrekidis uses the Newton-Raphson method to find a fixed point of a nonlinear mapping in a finite dimensional space that results from the discretization of the problem. His algorithm suffers from the undesirable convergence properties of Newton's method, but can work well provided the Jacobian exists. The method of Dieci does not identify the

---

1991 *Mathematics Subject Classification.* Primary 65L10, 34B15; Secondary 65S05.

cross-section of a 2-torus with its Poincaré hyperplane. Thus it is unable to follow the torus computation close to its breakdown.

In this paper we introduce the Hadamard graph transform approach to compute invariant circles. Each transform step involves solving a number of boundary value problems (BVP). One advantage of this approach is each BVP is independent of the others, and they can be solved concurrently. We illustrate the efficiency of the algorithm with two problems. The first example is a simple discrete map. The second is a differential equation for which the computed invariant circles are the cross-sections of a two-torus.

2. Formulation

The basic dynamical system we consider is that of the form

$$\dot{\mathbf{x}} = f(\mathbf{x}), \qquad \mathbf{x}(t_0) = \mathbf{x}_0, \tag{2.1}$$

with a solution denoted by $\phi_t(\mathbf{x}_0)$ and where $f : R^n \longrightarrow R^n$. Assume (2.1) has a periodic solution with period T and let $\mathbf{x}_0$ be a point on the periodic solution. Let Σ be an $n-1$ dimensional surface transverse to the vector field at the point $\mathbf{x}_0$ in an open set $V \subset \Sigma$ such that if $\phi_t(\mathbf{x})$ is C^r then trajectories in V return to Σ in time $\tau(\mathbf{x_0})$ close to T. The Poincaré map $P : V \to \Sigma$ is defined by

$$P(\mathbf{x}) := \phi_{\tau(\mathbf{x})}(\mathbf{x}) \in \Sigma, \qquad \mathbf{x} \in \mathbf{V}, \tag{2.2}$$

where $\tau(\mathbf{x})$ is the time of first return such that $\tau(\mathbf{x}_0) = T$ and $P(\mathbf{x}_0) = \mathbf{x}_0$ [W]. We assume that P is a homeomorphism. A fixed point of the Poincaré map corresponds to a periodic solution of the system (2.1) with period T. If (2.1) has an invariant two-torus near $\mathbf{x}_0$, then the Poincaré map possesses an invariant circle, ie., an invariant curve γ in R^n such that $P\gamma \subset \gamma$.

When considering the bifurcation of an ordinary differential system from a periodic solution to an invariant torus under the Poincaré map one can reduce it to a simpler bifurcation problem from a fixed point to an invariant circle. At the point of Naimark Sacker bifurcation, a plane Π containing an invariant curve can be determined from the eigenvectors of the two eigenvalues that cross the unit circle. The essential action of the system (2.1) takes place in this plane which actually gives a detailed asymptotic information about the system. Perturbation techniques can be used to compute the Euclidean projection of such invariant circles in this plane close to the bifurcation point in a bifurcation parameter space. There is a nonlinear coordinate transformation such that the plane Π is spanned by the first two coordinates in $\mathbb{R}^n$ [VV].

For some $\mathbf{x}_c \in \mathbb{R}^n, \theta \in S^1$ and $r \in \mathbb{R}^1$, we assume that the projection of the invariant curve γ in the plane Π, is given by

$$\gamma : \theta \in [0, 2\pi) \longrightarrow \gamma(\theta) = \mathbf{x}_c + (r(\theta)cos(\theta), r(\theta)sin(\theta), \mathbf{0})^T. \tag{2.3}$$

In the new coordinate system, equation (2.1) can be written as

$$\begin{pmatrix} \dot{\theta} \\ \dot{r} \\ \dot{\nu} \end{pmatrix} = \begin{pmatrix} f(\theta, r, \nu) \\ g(\theta, r, \nu) \\ h(\theta, r, \nu) \end{pmatrix}, \qquad 0 \leq t \leq \tau, \tag{2.4}$$

where $\nu \in \mathbb{R}^{n-2}$.

2.1. The Hadamard graph transform approach. This method is an application of the graph transform method developed in Fenichel [F]. It is a tool for finding (locally) attracting invariant manifolds. The approach works for both attractive and repellent curves but not for those with mixed attractivity. In this method one performs a sequence of graph transforms, where each graph transform itself requires the solution of finitely many BVP. We assume that the projection of the curve γ in (2.3) can be parameterized by

$$\gamma : \theta \longrightarrow \mathbf{R}(\theta), \qquad \theta \in [0, 2\pi), \qquad \mathbf{R} = \{R^1, R^2, \ldots, R^q\}. \tag{2.5}$$

We choose a mesh

$$\Omega_h := \{0 = \theta_1 \leq \theta_2 \leq \cdots \leq \theta_N = 2\pi\} \tag{2.6}$$

with $\mathbf{R}_i := \mathbf{R}(\theta_i)$, $i = 1, \ldots, N$ and let $\mathbf{R}^{old}(\theta)$ denote the initial approximation to γ. Let $\{P\mathbf{R}_i^{old}\}_{i=1}^N$ be the image of the points $\{\mathbf{R}_i^{old}\}_{i=1}^N$ under the Poincaré map P. If the curve γ is attracting (repelling) then the sequence of points $P^n\mathbf{R}_i$, $n = 1, ..., \infty(-\infty)$ for $i = 1, 2, ..., N$ gets closer to the curve γ. However, the sequence converges to a single point on γ if the curve contains a fixed point. Thus in this approach we update the points $\{\mathbf{R}_i^{old}\}_{i=1}^N$ after each Poincaré mapping.

We start by defining the projection operators U, V and W by

$$U(\theta, \mathbf{r}, \nu) := \theta, \qquad V(\theta, \mathbf{r}, \nu) := \mathbf{r}, \qquad W(\theta, \mathbf{r}, \nu) := \nu, \tag{2.7}$$

where $\theta \in S^1$, $\mathbf{r} \in \mathbb{R}^q$ and $\nu \in \mathbb{R}^{n-q-1}$. Define the Poincaré map P by

$$P(\theta^0, \mathbf{r}^0, \nu^0) := \phi_\tau(\theta^0, \mathbf{r}^0, \nu^0) = (\tilde{\theta}, \tilde{\mathbf{r}}, \nu^0) \tag{2.8}$$

where $\tau > 0$ and $\nu = \nu^0$ is an $(n - q - 1)$ dimensional hyperplane. Let $\mathbf{R}^{new} := H^\tau \mathbf{R}^{old}$ be the new approximation to $\mathbf{R}$ and H^τ be one Hadamard transformation. To get $\mathbf{R}^{new}$ we solve the boundary value problem $BVP(\bar{\theta}, \mathbf{R}^{old})$ given by

$$\begin{cases} \begin{pmatrix} \dot{\theta} \\ \dot{\mathbf{r}} \\ \dot{\nu} \end{pmatrix} = \begin{pmatrix} f(\theta, \mathbf{r}, \nu) \\ g(\theta, \mathbf{r}, \nu) \\ h(\theta, \mathbf{r}, \nu) \end{pmatrix}, & 0 \leq t \leq \tau, \\ \mathbf{r}(0) = \mathbf{R}^{old}, \ \theta(\tau) = \bar{\theta}, \ \nu(0) = \nu^0 = \nu(\tau) \end{cases} \tag{2.9}$$

to determine α and $\mathbf{R}^{new}(\bar{\theta})$ such that

$$U(\phi_\tau(\alpha, \mathbf{R}^{old}(\alpha), \nu^0)) = \bar{\theta}, \tag{2.10}$$

$$V(\phi_0(\alpha, \mathbf{R}^{old}(\alpha), \nu^0)) = \mathbf{R}^{old}(\alpha). \tag{2.11}$$

The Hadamard transform H^τ applied to $\mathbf{R}^{old}$ is defined by

$$(H^\tau \mathbf{R}^{old})(\bar{\theta}) := V(\phi_\tau(\alpha, \mathbf{R}^{old}(\alpha), \nu^0)) = \mathbf{R}^{new}(\bar{\theta}). \tag{2.12}$$

The approximation $\mathbf{R}$ gets very close to γ after some finite number of transformations. We solve the $BVP(\bar{\theta}, \mathbf{R}^{old})$ using simple and multiple shooting with either bisection or Newton's method. If the functions f, g, h and R^{old} are available locally then one can assign N $BVP(\bar{\theta}, \mathbf{R}^{old})s$ to N different processors.

3. Numerical examples

We apply our numerical scheme to two low dimensional problems to illustrate the robustness of the algorithm. The first example is the delayed logistic map and the second is a system of two coupled oscillators. In the second example the Poincaré map is not known explicitly and has to be determined during the numerical computation. The computation is done on a SUN SPARC 20 workstation and the 3-dimensional graphics produced on a Silicon Graphics Iris Indigo 2 Extreme.

3.1. Delayed logistic map. This is a population model that has been investigated in great details by Aronson et al [ACHM]. If we let N_n be the population density in the nth generation of a species and let a be a parameter reflecting the growth rate, then the population model is given by

$$N_{n+1} = aN_n(1.0 - N_{n-1}). \tag{3.1}$$

We set $x_n = N_{n-1}$ and $y_n = N_n$ to get

$$F_a(x_n, y_n) = (x_{n+1}, y_{n+1}) = (y_n, ay_n(1.0 - x_n)). \tag{3.2}$$

The map F_a has fixed points $(x^*, y^*) = (0,0)$ and $\frac{a-1}{a}(1,1)$. At $a = 2$ the fixed point $\frac{a-1}{a}(1,1)$ loses its stability and spawns an invariant circle via a Hopf bifurcation. It is known that for values of $a > 2.177$ the invariant curves are topologically circles but no longer differentiable [VV]. We are able to follow the circle up to $a = 2.270$ using a simple continuation in the parameter a. The result of the HGT method is shown in figure 1. Since the origin is a saddle, the outermost results show that we are approaching a homoclinic orbit. van Veldhuizen computes the invariant circles up to $a = 2.18$ and requires 1044 points to compute the invariant circle at $a = 2.18$ [VV], whereas our method requires 50 points to get similar results.

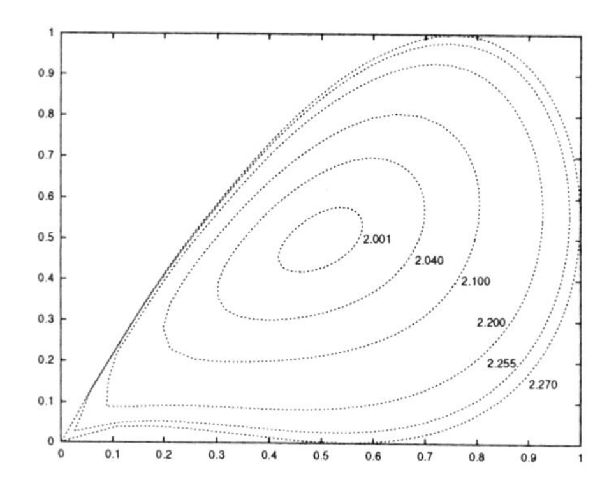

FIGURE 1. The invariant circles for the delayed logistic map with the value of a indicated. The vertical axis is y_n and the horizontal axis is x_n.

3.2. Coupled oscillators. In this example we look at the dynamics of two coupled planar oscillators which give rise to a system of ODEs in R^4. For the uncoupled system each oscillator has a unique periodic solution that is attracting and the coupled product system has a unique invariant torus that is also attracting. The torus persists for weak coupling when the coupling is linear and conservative.

However, the torus disappears for strong coupling. Our desire is to understand how the torus disappears.

The differential system is given by

$$(3.3)\qquad \begin{pmatrix} \dot{x}_1 \\ \dot{y}_1 \\ \dot{x}_2 \\ \dot{y}_2 \end{pmatrix} = \begin{pmatrix} \alpha_1 x_1 + \beta_1 y_1 - (x_1^2 + y_1^2)x_1 - \delta(x_1 + y_1 - x_2 - y_2) \\ \alpha_1 y_1 - \beta_1 x_1 - (x_1^2 + y_1^2)y_1 - \delta(x_1 + y_1 - x_2 - y_2) \\ \alpha_2 x_2 + \beta_2 y_2 - (x_2^2 + y_2^2)x_2 + \delta(x_1 + y_1 - x_2 - y_2) \\ \alpha_2 y_2 - \beta_2 x_2 - (x_2^2 + y_2^2)y_2 + \delta(x_1 + y_1 - x_2 - y_2) \end{pmatrix}.$$

With the parameterization $x_i = r_i cos(\theta_i)$ and $y_i = -r_i sin(\theta_i)$, $i = 1,2$ and $A = [sin(\theta_1 + \theta_2) - cos(\theta_1 - \theta_2)]$ we get

$$(3.4)\qquad \begin{pmatrix} \dot{\theta}_1 \\ \dot{\theta}_2 \\ \dot{r}_1 \\ \dot{r}_2 \end{pmatrix} = \begin{pmatrix} \beta_1 + \delta\{cos2\theta_1 - \frac{r_2}{r_1}[sin(\theta_1 - \theta_2) + cos(\theta_1 + \theta_2)]\} \\ \beta_2 + \delta\{cos2\theta_2 - \frac{r_1}{r_2}[sin(\theta_2 - \theta_1) + cos(\theta_1 + \theta_2)]\} \\ r_1(\alpha_1 - r_1^2) - \delta\{r_1(1 - sin2(\theta_1)) + Ar_2\} \\ r_2(\alpha_1 - r_2^2) - \delta\{r_2(1 - sin2(\theta_2)) + Ar_1\} \end{pmatrix}.$$

We let $\alpha_1 = \alpha_2 = 1.0$, $\beta_1 = \beta_2 = 0.55$ and the coupling parameter δ be our bifurcation parameter. The attracting limit circles of the uncoupled system are given by $x_i^2 + y_i^2 = \alpha_i \quad i = 1,2$. If we define the invariant torus by $M := (\theta_1, \theta_2, r(\theta_1,\theta_2), r_2(\theta_1,\theta_2))$ then the uncoupled system has an invariant torus $M = (\theta_1, \theta_2, 1, 1)$. We compute the Poincaré cross-section $\theta_1 = 0$ for different values of δ as shown in figures 2 and 3.

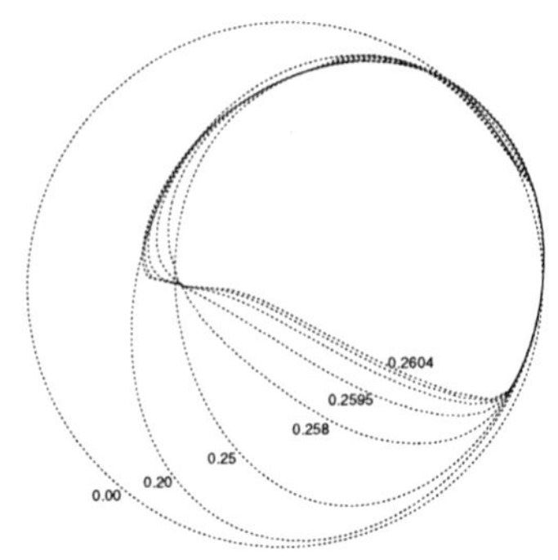

FIGURE 2. The cross-section $r_1(0, \theta_2)$ for different values of δ.

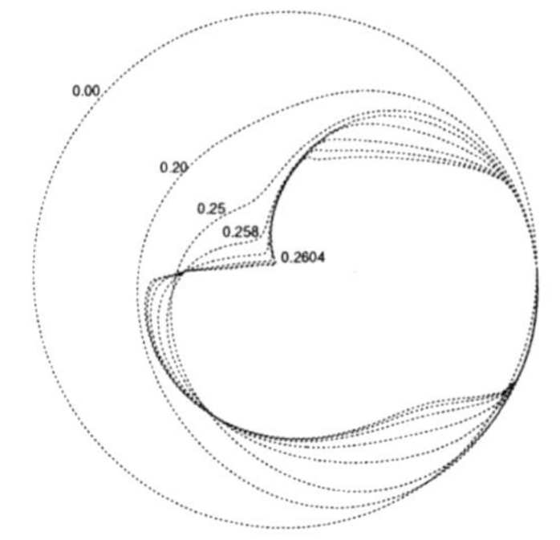

FIGURE 3. The cross-section $r_1(0, \theta_2)$ for different values of δ.

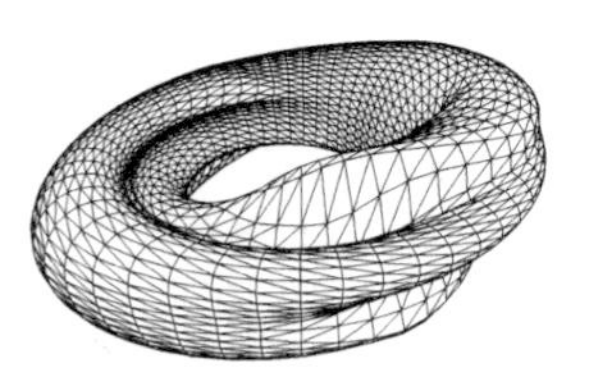

FIGURE 4. The torus $r_1(\theta_1, \theta_2)$ for $\delta = 0.26$.

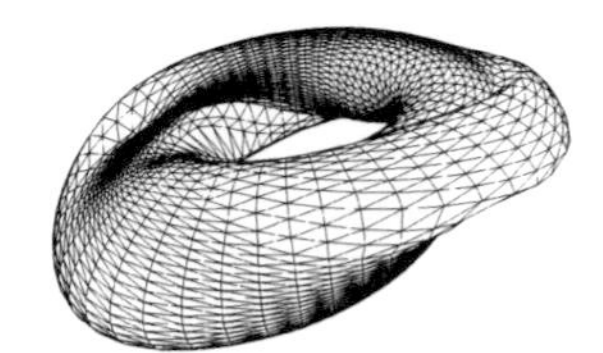

FIGURE 5. The torus $r_2(\theta_1, \theta_2)$ for $\delta = 0.26$.

During our computations from $\delta = 0$ to $\delta = 0.2600$ we use the initial guess $r_1 = 1$ and $r_2 = 1$. We use a simple continuation in δ with a stepsize of $\triangle\delta = 0.001$ from $\delta = 0.2600$ to $\delta = 0.2604$. We also use 50 mesh points to compute the cross-sections in figures 2 and 3. The results compare favorably with those of the pde approach of [ERS] but differ from those of [DB] for values of $\delta > 0.25$(about). Our results appear to be more reliable since they satisfy the symmetry properties of the differential equation. Figure 4 and 5 shows how deformed the torus is at $\delta = 0.26$.

4. Conclusion

We have been concerned with an algorithm to compute the invariant circles of maps. We focus on the invariant circle of Poincaré maps so as to compute the invariant two-tori. Our results are similar to those of van Veldhuizen [VV] and Dieci et al [DL]. Our algorithm does not restrict us to lower order piecewise interpolation schemes as in [VV]. We introduce an arc-length parametrization when a cusp begins to develop on the invariant circle. The algorithm works for only attracting or repelling invariant circles and not for those with mixed attractivity.

References

[ACHM] Aronson D.G., Chory M.A., Hall G.R. and McGehee R.P., *Bifurcation from an invariant circle for two parameter families of maps of the plane: a computer-assisted study, Comm. Math. Phys., 83, pp. 303-354, 1982.*

[DB] Dieci L. and Bader G., *Solution of the system associated with invariant tori approximation. II: multigrid methods, SIAM J. Sci. Comp., 15, pp. 1375-1400, 1994.*

[DL] Dieci L. and Lorenz J., *Computation of invariant tori by the method of characteristics, SIAM J. Numer. Anal., 32, pp. 1436-1474, 1995.*

[ERS] Edoh K.D., Russell R.D. and Sun W., *Numerical approximation of invariant tori using orthogonal collocation. In preparation.*

[F] Fenichel N., *Persistence and smoothness of invariant manifolds for flows, Indiana Univ. Math. J., 21, pp. 193-226, 1971.*

[KASP] Kevrekidis I.G., Aris R., Schmidt L.D. and Pelican S., *Numerical computation of invariant circles of maps. Physica 16D, pp. 243-251, 1985.*

[VV] Van Veldhuizen M., *A new algorithm for the numerical approximation of an invariant curve, SIAM J Sci. Stat. Comp. 8, pp. 951-962, 1987.*

[W] Wiggins S., *An introduction to Applied Nonlinear Dynamical Systems and Chaos, Springer Verlag, New York Inc, 1990.*

Department of Mathematics & Statistics, Simon Fraser University, Burnaby B.C. V5A 1S6, Canada

Current address: Department of Mathematics & Computer Science, Elizabeth City State University, Elizabeth City, N.C. 27909 U.S.A.

E-mail address: edoh@cs.sfu.ca

DIMACS Series in Discrete Mathematics
and Theoretical Computer Science
Volume **34**, 1997

CLASSIFICATION OF NILPOTENT ORBITS IN SYMMETRIC SPACES

ALFRED G. NOËL

ABSTRACT. We present a new classification of nilpotent orbits of real reductive Lie algebras under the action of their adjoint group. This classification generalizes the one given by P. Bala and R. Carter in 1976, for complex semisimple Lie algebras.[1],[2]

1. Introduction

Let $\mathfrak{g}$ be a real reductive Lie algebra with adjoint group G and $\mathfrak{g}_{\mathbb{C}}$ its complexification. Also let $\mathfrak{g} = \mathfrak{k} \oplus \mathfrak{p}$ be the corresponding Cartan decomposition of $\mathfrak{g}$. Finally, let θ be a Cartan involution of $\mathfrak{g}$ and σ be the conjugation of $\mathfrak{g}_{\mathbb{C}}$ with regard to $\mathfrak{g}$. Then, $\mathfrak{g}_{\mathbb{C}} = \mathfrak{k}_{\mathbb{C}} \oplus \mathfrak{p}_{\mathbb{C}}$ where $\mathfrak{k}_{\mathbb{C}}$ and $\mathfrak{p}_{\mathbb{C}}$ are obtained by complexifying $\mathfrak{k}$ and $\mathfrak{p}$ respectively. Denote by $K_{\mathbb{C}}$ the connected subgroup of the adjoint group $G_{\mathbb{C}}$ of $\mathfrak{g}_{\mathbb{C}}$, with Lie algebra $\mathfrak{k}_{\mathbb{C}}$. We prove that the orbits $K_{\mathbb{C}}.e$ are in one-to-one correspondence with the triples of the form $(l, q_l, \mathfrak{w})$, where e is a non zero nilpotent in $\mathfrak{p}_{\mathbb{C}}$, l is a minimal (θ, σ)-stable Levi subalgebra of $\mathfrak{g}_{\mathbb{C}}$ containing e, q_l is a θ stable parabolic subalgebra of $[l, l]$ and $\mathfrak{w}$ is a certain $L \cap K_{\mathbb{C}}$ prehomogeneous subspace of $q_l \cap \mathfrak{p}_{\mathbb{C}}$ containing e. L is the connected subgroup of $G_{\mathbb{C}}$ with Lie algebra l. Thus, we obtain a classification for real nilpotents G-orbits in $\mathfrak{g}$ via the Kostant-Sekiguchi correspondence.[7]

2. The Classification

Let $\mathfrak{g}_0 = \mathfrak{k} \oplus i\mathfrak{p}$. Then, $\mathfrak{g}_0$ is a compact real form of $\mathfrak{g}_{\mathbb{C}}$ and $\mathfrak{g}_{\mathbb{C}} = \mathfrak{g}_0 \oplus i\mathfrak{g}_0$. Let κ be the Killing form on $\mathfrak{g}_{\mathbb{C}}$ and τ, the conjugation of $\mathfrak{g}_{\mathbb{C}}$ with respect to $\mathfrak{g}_0$. For X and Y in $\mathfrak{g}_{\mathbb{C}}$ define $\kappa'(X, Y) = -\kappa(X, \tau(Y))$. Then κ' is a hermitian form on $\mathfrak{g}_{\mathbb{C}}$.

Let $\mathfrak{q} = l \oplus u$ be a θ-stable parabolic subalgebra of $\mathfrak{g}_{\mathbb{C}}$. Let $\mathfrak{m}$ be the orhogonal complement of $[u \cap \mathfrak{k}_{\mathbb{C}}, [u \cap \mathfrak{k}_{\mathbb{C}}, u \cap \mathfrak{p}_{\mathbb{C}}]]$ relative to κ' inside $u \cap \mathfrak{p}_{\mathbb{C}}$. Define $\mathfrak{w}$ to be an $L \cap K_{\mathbb{C}}$ module in $\mathfrak{m}$. Finally, let $\hat{\mathfrak{w}} = \mathfrak{w} \oplus [l \cap \mathfrak{p}_{\mathbb{C}}, \mathfrak{w}]$ be an L-module. Clearly, $\hat{\mathfrak{w}}$ is θ-stable.

1991 *Mathematics Subject Classification.* 17B20,17B70.
Key words and phrases. parabolic subalgebras, nilpotent orbits, reductive Lie algebras.
The author thanks his advisor, Donald R. King, for his helpful suggestions.

Definition. Define $\mathfrak{L}$ to be the set of triples $\{\mathfrak{g}_{\mathbb{C}}, \mathfrak{q}, \mathfrak{w}\}$ such that:

1. $\mathfrak{w}$ has a dense $L \cap K_{\mathbb{C}}$ orbit.
2. dim $l \cap \mathfrak{k}_{\mathbb{C}} =$ dim $\mathfrak{w}$.
3. L has an open dense orbit on $\hat{\mathfrak{w}}$ and that dense orbit comes from an element $\hat{e} \in \mathfrak{w}$ such that $(L \cap K_{\mathbb{C}}).\hat{e}$ is dense in $\mathfrak{w}$.
4. $\hat{\mathfrak{w}} \perp [u, [u, u]]$
5. $[u, \hat{\mathfrak{w}}] \cap \hat{\mathfrak{w}} = \{0\}$

Let e be a non-zero nilpotent element of $\mathfrak{p}_{\mathbb{C}}$. It is known that e can be embedded in a normal triple (x, e, f) in Kostant-Rallis's sense that is $[x, e] = 2e$, $[x, f] = -2f$ and $[e, f] = x$, with $x \in \mathfrak{k}_{\mathbb{C}}$, e and $f \in \mathfrak{p}_{\mathbb{C}}$ [4]. Since we are interested in $K_{\mathbb{C}}$-conjugacy classes of nilpotents in $\mathfrak{p}_{\mathbb{C}}$ we can assume that $\sigma(e) = f$.[7]

From the representation theory of $\mathfrak{sl}_2(\mathbb{C})$, $\mathfrak{g}_{\mathbb{C}}$ has the following eigenspace decomposition:

$$\mathfrak{g}_{\mathbb{C}} = \bigoplus_{j \in \mathbb{Z}} \mathfrak{g}_{\mathbb{C}}^{(j)} \text{ where } \mathfrak{g}_{\mathbb{C}}^{(j)} = \{z \in \mathfrak{g}_{\mathbb{C}} | [x, z] = jz\}.$$

The θ-stable subalgebra $\mathfrak{q}_x = \bigoplus_{j \in \mathbb{N}} \mathfrak{g}_{\mathbb{C}}^{(j)}$ is a parabolic subalgebra of $\mathfrak{g}_{\mathbb{C}}$ with a Levi part $l = \mathfrak{g}_{\mathbb{C}}^{(0)}$ and nilradical $u = \bigoplus_{j \in \mathbb{N}^*} \mathfrak{g}_{\mathbb{C}}^{(j)}$. Call $\mathfrak{q}_x$ the Jacobson-Morosov parabolic subalgebra of e relative to the triple (x, e, f).

Definition. A nilpotent element e in $\mathfrak{p}_{\mathbb{C}}$ (or its $K_{\mathbb{C}}$-orbit) is *noticed* if the only (σ, θ)-stable Levi subalgebra of $\mathfrak{g}_{\mathbb{C}}$ containing e (or equivalently meeting $K_{\mathbb{C}}.e$) is $\mathfrak{g}_{\mathbb{C}}$ itself.

A Levi subalgebra l contains e if and only if $[l, l]$ does. Thus if e is noticed in l it is actually noticed in the semi-simple subalgebra $[l, l]$. Finally if l is (θ, σ)-stable then so is $[l, l]$, and any nilpotent $e \in \mathfrak{p}_{\mathbb{C}}$ is noticed in any minimal (θ, σ)-stable Levi subalgebra l containing it. If e is noticed then the normal triple (x, e, f) is said to be noticed.

Let $\mathfrak{S}$ be the set of noticed normal triples (x, e, f) of $\mathfrak{g}_{\mathbb{C}}$.

We have a map $\mathfrak{F}$ from $\mathfrak{S}$ to $\mathfrak{L}$ which associates a triple (x, e, f) of $\mathfrak{S}$ to an element $(\mathfrak{g}_{\mathbb{C}}, \mathfrak{q}_x, \mathfrak{w})$ of $\mathfrak{L}$ where $\mathfrak{q}_x$ is the θ-stable parabolic subalgebra relative to (x, e, f) and $\mathfrak{w} = \mathfrak{g}_{\mathbb{C}}^2 \cap \mathfrak{p}_{\mathbb{C}}$. From Kostant and Rallis [4] we know that $L.e$ (respectively $L \cap K_{\mathbb{C}}.e$) is dense on $\mathfrak{g}_{\mathbb{C}}^{(2)}$ (respectively $\mathfrak{g}_{\mathbb{C}}^{(2)} \cap \mathfrak{p}_{\mathbb{C}}$). Also dim $\mathfrak{g}_{\mathbb{C}}^{(0)} \cap \mathfrak{k}_{\mathbb{C}} =$ dim $\mathfrak{g}_{\mathbb{C}}^{(2)} \cap \mathfrak{p}_{\mathbb{C}}$ because e is noticed. Since $[\mathfrak{p}_{\mathbb{C}}^x, \mathfrak{g}_{\mathbb{C}}^{(2)} \cap \mathfrak{p}_{\mathbb{C}}] = \mathfrak{g}_{\mathbb{C}}^{(2)} \cap \mathfrak{k}_{\mathbb{C}}$ [6], it is clear that $\mathfrak{F}$ is well defined.

From a theorem of Kostant & Rallis [4], there is a bijection between the non-zero nilpotent $K_{\mathbb{C}}$-orbits in $\mathfrak{p}_{\mathbb{C}}$ and the $K_{\mathbb{C}}$-conjugacy classes of normal triples. Two normal noticed triples (x, e, f) and (x', e', f') are $K_{\mathbb{C}}$ conjugate if and only if their corresponding triples $(\mathfrak{g}_{\mathbb{C}}, \mathfrak{q}, \mathfrak{w})$ and $(\mathfrak{g}_{\mathbb{C}}, \mathfrak{q}', \mathfrak{w}')$ are $K_{\mathbb{C}}$ conjugate. Hence, $\mathfrak{F}$ induces a one to one map from $K_{\mathbb{C}}$-orbits of $\mathfrak{S}$ and the $K_{\mathbb{C}}$-conjugacy classes of the triples of $\mathfrak{L}$. The following theorem tells us that such a map is also surjective.

Theorem 1. *For any triple $\{\mathfrak{g}_{\mathbb{C}}, \mathfrak{q}, \mathfrak{w}\}$ of $\mathfrak{L}$ there exists a normal triple (x, e, f) in $\mathfrak{S}$ such that $\mathfrak{q}$ is the Jacobson-Morosov parabolic subalgebra for (x, e, f) and $\mathfrak{w} = \mathfrak{g}_{\mathbb{C}}^{(2)} \cap \mathfrak{p}_{\mathbb{C}}$.*

Proof. See [6]. ■

Let l be a (θ, σ)-stable Levi Subalgebra of $\mathfrak{g}_{\mathbb{C}}$. Define the set of triples $(l, \mathfrak{q}_l, \mathfrak{w}_l)$ to have the same properties as the triples of $\mathfrak{L}$, replacing $\mathfrak{g}_{\mathbb{C}}$ by l. Here $\mathfrak{q}_l$ is a θ-stable parabolic subalgebra of $[l, l]$. Then we have:

Theorem 2. *There is a one-to-one correspondance between nilpotent $K_{\mathbb{C}}$-orbits on $\mathfrak{p}_{\mathbb{C}}$ and $K_{\mathbb{C}}$-conjugacy classes of triples $(l, \mathfrak{q}_l, \mathfrak{w}_l)$.*

Proof. We noted before that a Levi subalgebra l contains a nilpotent element $e \in \mathfrak{p}_{\mathbb{C}}$ if and only if $[l, l]$ does. Two Levi subalgebras are $K_{\mathbb{C}}$-conjugate if and only if their derived subalgebras are. Each nilpotent $e \in \mathfrak{p}_{\mathbb{C}}$ can be put in a normal triple (x, e, f) inside of the minimal Levi subalgebra l containing e. Any two minimal Levi subalgebras containing e are conjugate under $K_{\mathbb{C}}^e$ [6]. Hence the theorem follows from theorem 1. ■

3. Example

Let $\mathfrak{g}$ be $\mathfrak{sl}(3, \mathbb{R})$, the set of 3×3 real matrices of trace 0, Then $\mathfrak{g}_{\mathbb{C}} = \mathfrak{sl}(3, \mathbb{C})$, $\mathfrak{k}_{\mathbb{C}} = \mathfrak{so}(3, \mathbb{C})$, and $\mathfrak{p}_{\mathbb{C}}$ is the space of 3×3 complex symmetric matrices. The Cartan involution θ is defined as $\theta(X) = -X^T$ for $X \in \mathfrak{g}$. Denote by $\bar{Y}$, the complex conjugate of a matrix $Y \in \mathfrak{g}_{\mathbb{C}}$.

The set of orthogonal matrices ($K_{\mathbb{C}}$) preserves the set of symmetric matrices ($\mathfrak{p}_{\mathbb{C}}$) under conjugation. The nilpotent orbits of $K_{\mathbb{C}}$ on $\mathfrak{p}_{\mathbb{C}}$ are parametrized by the partitions of 3. Therefore, there are exactly two non zero nilpotent classes since the zero nilpotent class corresponds to the partition $[1, 1, 1]$. A computation shows that the following matrices

$$H_1 = \begin{pmatrix} 0 & i & 0 \\ -i & 0 & 0 \\ 0 & 0 & 0 \end{pmatrix}, H_2 = \begin{pmatrix} 1 & 0 & 0 \\ 0 & 1 & 0 \\ 0 & 0 & -2 \end{pmatrix},$$

$$E_1 = \frac{1}{2}\begin{pmatrix} i & 1 & 0 \\ 1 & -i & 0 \\ 0 & 0 & 0 \end{pmatrix}, E_2 = \begin{pmatrix} 0 & 0 & i \\ 0 & 0 & 1 \\ i & 1 & 0 \end{pmatrix}, E_3 = \begin{pmatrix} 0 & 0 & -i \\ 0 & 0 & -1 \\ i & 1 & 0 \end{pmatrix}$$

generate the only θ-stable Borel subalgebra $\mathfrak{b}$ of $\mathfrak{g}_{\mathbb{C}}$. Let

$$\mathfrak{b} = \mathbb{C}H_1 \oplus \mathbb{C}E_3 \oplus \mathbb{C}H_2 \oplus \mathbb{C}E_1 \oplus \mathbb{C}E_2.$$

Of course $\mathfrak{b}$ is conjugate to the set of upper triangular matrices of $\mathfrak{sl}(3, \mathbb{C})$.

We see that $\mathfrak{l} \cap \mathfrak{k}_{\mathbb{C}} = \mathbb{C}H_1$ and $\mathfrak{m} = \mathbb{C}E_1 \oplus \mathbb{C}E_2$. Let $\mathfrak{w}_1 = \mathbb{C}E_1$ and $\mathfrak{w}_2 = \mathbb{C}E_2$. Clearly, $L \cap K_{\mathbb{C}}$ has a dense orbit on $\mathfrak{w}_1$ and $\mathfrak{w}_2$ respectively for $[H_1, E_1] = 2E_1$ and $[H_1, E_2] = E_2$. Also

$$\dim \mathfrak{l} \cap \mathfrak{k}_{\mathbb{C}} = \dim \mathfrak{w}_1 = \dim \mathfrak{w}_2 = 1.$$

For each $\mathfrak{w}_i$ one verifies easily that the triple ,$(\mathfrak{g}_{\mathbb{C}}, \mathfrak{b}, \mathfrak{w}_i)$, satisfies all the requirements specified above.

Thus we obtain the following correspondence:

$$(H_1, E_1, \bar{E}_1) \longleftrightarrow (\mathfrak{g}_{\mathbb{C}}, \mathfrak{b}, \mathfrak{w}_1)$$

$$(2H_1, E_2, \bar{E}_2) \longleftrightarrow (\mathfrak{g}_{\mathbb{C}}, \mathfrak{b}, \mathfrak{w}_2)$$

4. Summary

Nilpotent orbits have been used extensively in Representation Theory. Their geometric structure is still being investigated by several researchers. Θ-stable parabolic subalgebras, also, play an important role in Representation Theory, specifically through the work of Zuckerman and Vogan on Cohomological Induction. Our classification relates the two concepts.

References

1. P. Bala and R. Carter. *Classes of unipotent elements in simple algebraic groups I*, Math. Proc. Camb. Phil. Soc. **79** (1976), 401-425.

2. P. Bala and R. Carter. *Classes of unipotent elements in simple algebraic groups II*, Math. Proc. Camb. Phil. Soc. **80** (1976), 1-18.

3. D. H. Collingwood, W. M. McGovern. *Nilpotent Orbits in Semisimple Lie Algebras.* Van Nostrand Reihnhold Mathematics Series, New York 1993.

4. B. Kostant, S. Rallis. *Orbits and Representations associated with symmetric spaces.* Amer. J. Math. **93**(1971), 753-809.

5. A. W. Kapp and D. A. Vogan. *Cohomological Induction and Unitary Representations.* **45**, Princeton New Jersey, 1995.

6. A. G. Noël. *Ph.D. Thesis.* Northeastern University, Boston (in preparation)

7. J. Sekiguchi. *Remarks on a real nilpotent orbits of a symmetric pair.* J. Math. Soc. Japan. Vol. **39**, No. 1, 1987, 127-138.

8. D. A. Vogan. *Representation of Real Reductive Lie Groups.* Birkhäuser, Progress in Mathematics Boston, 1981.

DEPARTMENT OF MATHEMATICS, NORTHEASTERN UNIVERSITY, BOSTON, MA, 02115, USA
E-mail address: anoel@lynx.neu.edu

DIMACS Series in Discrete Mathematics
and Theoretical Computer Science
Volume **34**, 1997

Evaluating Texture Measures for Low-Level Features in Color Images of Human Skin

Kori E. Needham

ABSTRACT. The investigation undertaken in this research project is based on two processes of image manipulation, namely image processing and computer vision, which constitute the foundation upon which texture measures have been developed. Within the scope of this project, several color texture measures will be compared. Two of these measures are discussed in some detail here to explain their potential effectiveness in unsupervised comparisons of low-level features in high-quality 24-bit color images of human skin captured from facial cheeks.

1. Introduction

1.1. Background. In the past, surgeons could do very little to help their patients to fully recover from scars caused by lacerations, frost-bite, disease, burns, etc. Typically, all they could do was to apply ointments and lotions to soothe the region of scarred tissue and then refer these patients to psychiatrists for help in coping with living with permanently scarred skin. With the advent of modern technology and new

1991 *Mathematics Subject Classification.* Primary 68U10, 62–07, 15A18.

surgical techniques, surgeons have been able to perform skin grafts and apply cosmetics to help make the scarred region more visually pleasing to the patient and less of an attention-grabber for others. Unfortunately, assessing the success of these methods of treatment to make the region appear *better* has been extremely subjective and somewhat ambiguous. *Better* is based only on subjective interpretations of which method of treatment (cosmetics, surgery, etc.) makes the scarred tissue more closely resemble the patient's normal skin. A more objective method of evaluating the effectiveness of a surgeon's treatment of the scar is necessary. This method should be superior to the current method, because it would utilize the basic underlying characteristics or features (i.e. tone, coarseness, color, etc.) of regions of skin being compared in order to assess whether or not the treatment was successful in making the scarred region look *better*.

Currently, at the University of South Florida's Tampa campus, doctors and researchers in the Burn Center and the Psychiatry Department are interested in treatment of burn victims. They are researching ways to make burn scars actually vanish or at least make them look significantly *better*. They have questioned whether computer vision and image processing techniques could be used to objectively evaluate the results of a particular surgeon's method of treating a burn scar. Specifically, this research seeks to determine which computable texture measures are suitable for use with color images of skin, and which of these texture measures agree well with human judgements about the similarity of the features of one patch of skin with that of another patch. Whichever texture measure is chosen, it must work for a variety of skin tones, for example, pale, tan, light brown, black.

1.2. Objectives. Within The Burn Scar Tracking Project, there are four major phases which are associated with the development of the software for analyzing the burn scars. Phase one is to identify some suitable textures measures which can classify, or categorize, normal unblemished human skin images from a particular region of the body. Phase two is to modify these measures by changing or weighting some of the parameters in the correlation functions to work with more diverse images from various parts of the body. Phase three involves acquiring from hospitals and/or medical institutions high-quality 24-bit color images of regions of the human skin containing burn scars. The final part of the project, phase four, involves modifying the graphical user interface and the software program overall so that a stable version of it can be tested in an actual hospital setting. This research will deal with the initial phase of the project described above. The goal of this investigation is to find and implement a texture measure which will work on color images of human skin. In addition, the measure should be invariant to small perturbations of lighting conditions and to small deviations in the location of the viewpoint. And last, but not least, the measure needs to be capable of recognizing the variations in low level features which make one image similar to or different from another image in the database.

Once this texture measure is identified and verified, it would be utilized in step three (below) of a hypothetical protocol which a surgeon might use in the process of treating the scar of a burn victim. In step one, a nurse would photograph the scar area with a 24-bit digital camera and then transmit the image data to an image processing and analysis software package. This software would subsequently identify the scar region more precisely by segmenting it away from the surrounding unblemished skin. Step two follows some pre-designated period of time (days, weeks, or months depending on the severity of the scar) during which the surgeon would have attempted to treat the scar using one specific method or a combination of any number of methods. At this time, a nurse would again photograph the scar area along with an area of normal skin from a similar part of the body and transmit both images to the same scar analysis software package. During step three, the current scar image would be automatically compared to both the previous scar image and the one taken of the normal region of skin, thereby classifying the scar as looking better, the same, or worse. Each classification is based on a distance value which is an indirect result of a correlation function being performed on each pixel in the images being compared.

Table 1. Classification of A Burn Scar After Treatment.

	Range of Distance Values Relative to Normal Skin	
Classification	**Previous Scar Image**	**Current Scar Image**
better	**D**	**(D - x)**
same	**D**	**D**
worse	**D**	**(D + x)**

x = magnitude of the change in distance values from previous scar to current scar. Refer to Figure 1 for a more detailed representation of distance values.

By closely scrutinizing the distance values which would directly correspond to the development of the burn scar, both the patient and the surgeon would be better equipped to determine which step to take next in the treatment process. Consequently, if further treatment is deemed necessary, the photography and analysis process would be repeated by the nurse.

2. Two Promising Texture Measures

Based on the reading about various segmentation and classification methods, a hypothesis has been made about which texture measures have the most potential for successfully matching color images of human skin. Since the test database is composed of images of cheeks taken from five different viewpoints purposely chosen to cause deviations in lighting, the measures chosen were those whose designers claim they are relatively invariant to these parameters. By the process of elimination, two promising texture measures were identified. Although both will be explained in detail below, only the first one has undergone preliminary testing for its accuracy in classifying of color images of human skin. The second one should inevitably be tested by another member of the Burn Scar Tracking Project. The first measure has already been successful in classifying color images of identical scenes with variations in lighting, while the second measure has achieved success in classifying color images of identical scenes taken from a variety of viewpoint locations. Below are the texture measures that were selected.

2.1. Illumination-Invariant Spatial Correlations For Color Textures. The illumination-invariant texture measure [1] uses the intensities at a given pixel in an image to derive the sample covariance formula

$$R_{ij}(n,m) = E[(I_i(a,b) - X_i)(I_j(a+n, b+m) - X_j)]. \quad (1)$$

Here, **a** and **b** represent the location of a pixel in an image, **n** and **m** represent a shift or translation about the pixel at (a,b), $\mathbf{I_i}$ represents the three (red, green, blue) spectral intensities for each pixel in the image, $\mathbf{X_i}$ represents the mean intensity of an image, **E** represents the Expected Value, and $\mathbf{R_{ij}}$ represents the six correlations of spectral intensities between two pixels. These six distinct correlation functions include three autocorrelations and three cross correlations (i.e. R_{rr}, R_{rg}, R_{rb}, R_{gg}, R_{gb}, R_{bb}), each of which are normalized by dividing by $\mathbf{R_{rr}(0,0)}$ to eliminate effects that might be caused by illumination intensity.

Given P=n*m, these correlation functions are used to construct a Px6 correlation matrix

$$C = [\ c_1, c_2, \ldots, c_6\ ],$$

where $\mathbf{c_j}$ represents a vector of one of the six correlation functions performed on all the pixels in an image. Therefore, the length of $\mathbf{c_j}$ is equal to p(p-2)/2, where p is the number of pixels in the image. **C** is characterized by an orthonormal basis computed from the singular value decomposition (SVD) given by

$$C = U\Sigma V^T \tag{2}$$

where the columns of the P x 6 matrix $\mathbf{U} = [u_1, u_2, \ldots, u_6]$ are orthonormal eigenvectors of CC^T, Σ is a 6x6 diagonal matrix of singular values ($\sigma_1, \sigma_2, \ldots, \sigma_6$), and the columns of the 6 x 6 matrix $\mathbf{V} = [v_1, v_2, \ldots, v_6]$ are orthonormal vectors of CC^T. If C' is related to C by a linear transformation, then we can use the basis vectors ($u_1, u_2, \ldots, u_6$) which correspond to C in order to represent the columns ($\mathbf{m_j}$) of C' with the equation

$$m_j = (u_j^T\, c_i')u_j. \tag{3}$$

Finally, the best approximation to the columns of C' using the basis ($u_1, u_2, \ldots, u_6$) results in the derivation of the distance formula

$$D = \Sigma_i \,||\, c_i' - (m_1 + m_2 + \ldots + m_6) \,||^2, \tag{4}$$

for $1 \leq i \leq 6$ where **D**=0 represents a perfect match, or in other words, C' is in fact a linear transformation of C. Furthermore, given a color texture represented by C, other color textures represented by C' and also related to C by a variation in illumination will yield small distance values D.

2.2. Geometry-Invariant Spatial Correlations For Color Textures. Similar to the illumination-invariant measure described above, the geometry-invariant measure [2] is based on the covariance formula which yields the same six correlation functions

$$R_{ij}(n,m) = E[(I_i(a,b) - X_i)\,(\,I_j(a+n, b+m) - X_j\,)],$$

discussed in the previous section, and which are computed from the intensities at each pixel in a given image. In order to make these correlation functions, $R_{ij}(n,m)$, independent of scaling factors such as the incident light intensity, they are also normalized by dividing by $R_{rr}(0,0)$. Two image textures are then compared and said to be from the same facial cheek of a particular subject if the distance equation

$$D_{ij} = \Sigma_m \Sigma_n [R_{ij}(\mathbf{M}_2^{-1} \eta) - R_{ij}'(\eta)]^2,$$

gives $\mathbf{D}_{ij}=0$ for each of the six correlation functions. Here, η is the vector [n, m] and $\mathbf{M}_2$ is the image transformation matrix used to relate a plane's image-correlation functions with the correlation functions for a transformed version of the plane in

$$R_{ij}(\mathbf{M}_2^{-1} \eta) = R_{ij}'(\eta).$$

Since the correlations are estimates, we will generally be unable to transform exactly from one set of correlation functions to another. Therefore, simultaneously minimizing the squared difference of the six correlations yields a new distance equation

$$D_{ij} = \Sigma_{ij} \{\Sigma_m \Sigma_n [R_{ij\,\mathrm{estimate}}(\mathbf{M}_2^{-1} \eta) - R_{ij0}(\eta)]^2 \}.$$

Consequently, with this texture measure, a successful image classification occurs when the value of $\mathbf{D}_{ij}$ is less than a specified threshold which is usually determined after some experimentation.

3. Conclusions and Proposal For Future Research

3.1. Conclusions. This research project investigated numerous texture measures and evaluated the effectiveness of an illumination-invariant texture measure on color images of human skin captured from facial cheeks. The experimental results show the classification of the images was not achieved due to inaccurate distance values. It is suggested that the illumination-invariant measure be re-implemented with another SVD algorithm. In essence, the results of this project were not as positive as the author had hypothesized. Nevertheless, the research undertaken was an excellent learning experience for the author in both the fields of Image Processing and Computer Vision.

3.2. Proposal For Future Research. Future research related to this project would need to incorporate the acquisition of approximately 600-700 new images (containing a variety of low-level features) which could be added to the current database of 520 images and then scrutinized according to the guidelines of a diversity control protocol. Then the geometry-invariant texture measure described in Section 2.2, as well as a few more color texture measures, should be implemented on the larger, more diverse database (following the guidelines of Phase one of the Burn Scar Tracking Project) to test whether these new measures could successfully match images of human cheeks from the same subject's face. Furthermore, once Phase one is successfully completed, Phases two and three need to be implemented. At this point, the current set of database images should be augmented to include (or replaced by) images of scars from the case histories of burn victims. Therefore, the measures could be evaluated for their effectiveness in classifying a scar throughout a particular treatment process. Hopefully, this endeavor is only the beginning of a very lucrative, interesting, and educational research experience in the fields of Medical Imaging, Image processing, and Computer Vision.

4. References

[1] Healey, Glenn, and Lizhi Wang. 1995. The Illumination-Invariant Recognition of Color Texture. Fifth International Conference on Computer Vision. Longer version published in Journal of the Optical Society of America A. November.

[2] Kondepudy, Raghava, and Glenn Healey. 1994. Use of invariants for recognition of three-dimensional color textures. Journal of the Optical Society of America A. June. Pp. 3037-3049.

Needham, Kori E. 1996. Evaluating Texture Measures for Comparisons of Low-Level Features in Color Images of Human Skin. Eckerd College Senior Thesis. May.

Department of Computer Science, University of North Carolina, Chapel Hill, NC 27599-3175.

needham@cs.unc.edu

DIMACS Series in Discrete Mathematics
and Theoretical Computer Science
Volume **34**, 1997

Lattice Paths and RNA Secondary Structures

ASAMOAH NKWANTA

ABSTRACT. Four infinite lower-triangular matrices, each of whose entries count lattice paths or random walks, are presented and denoted as P, C_0, C and M where M is a Motzkin triangle, C and C_0 are Catalan triangles, and P is Pascal's triangle. By matrix multiplication, another infinite lower-triangular matrix denoted as R is defined by $C_0 \cdot R = M$. Then, the following results are proved about R:

1) $M \cdot R = C$.
2) The entries in the left most column of R count the number of RNA secondary structures of length n (from molecular biology).
3) A combinatorial interpretation of R is given in terms of lattice paths.
4) There is a one-to-one correspondence between RNA secondary structures and these lattice paths.
5) The first moments of R are every other Fibonacci number.

Results related to the Narayana numbers and noncrossing partitions are also discussed.

1. Introduction

We consider four infinite lower-triangular matrices denoted as C_0, M, C and P, where the first few terms of each triangle are listed below:

$$C_0 = \begin{bmatrix} 1 & & & & & \\ 0 & 1 & & & & \\ 1 & 0 & 1 & & & \\ 0 & 2 & 0 & 1 & & \\ 2 & 0 & 3 & 0 & 1 & \\ \vdots & \vdots & \vdots & \vdots & \vdots & \ddots \end{bmatrix}, \qquad M = \begin{bmatrix} 1 & & & & & \\ 1 & 1 & & & & \\ 2 & 2 & 1 & & & \\ 4 & 5 & 3 & 1 & & \\ 9 & 12 & 9 & 4 & 1 & \\ \vdots & \vdots & \vdots & \vdots & \vdots & \ddots \end{bmatrix},$$

$$C = \begin{bmatrix} 1 & & & & & \\ 2 & 1 & & & & \\ 5 & 4 & 1 & & & \\ 14 & 14 & 6 & 1 & & \\ 42 & 48 & 27 & 8 & 1 & \\ \vdots & \vdots & \vdots & \vdots & \vdots & \ddots \end{bmatrix}, \quad \text{and} \quad P = \begin{bmatrix} 1 & & & & & \\ 1 & 1 & & & & \\ 1 & 2 & 1 & & & \\ 1 & 3 & 3 & 1 & & \\ 1 & 4 & 6 & 4 & 1 & \\ \vdots & \vdots & \vdots & \vdots & \vdots & \ddots \end{bmatrix}.$$

1991 *Mathematics Subject Classification.* Primary 05A15; Secondary 92D20.

Partially supported by an HBCU fellowship from the Jet Propulsion Laboratory, Pasadena, CA.

P is the well known Pascal triangle. M is called a Motzkin triangle since its left most column contains the Motzkin numbers, $m_n = \sum_{k\geq 0} \frac{1}{k+1}\binom{2k}{k}\binom{n}{2k}$. Similarly, C_0 and C are called Catalan triangles since their left most columns contain the Catalan numbers, $c_n = \frac{1}{n+1}\binom{2n}{n}$. The entries of C_0 are sometimes referred to as the aerated Catalan numbers since zeros are between each number. These triangles often arise in combinatorial applications. For instance, the Motzkin triangle has interpretations as random walks [4], and as interval graphs [8]. The Catalan triangles have interpretations as ballot sequences [17], and as lattice paths (or walks) with various restrictions [6], [7]. Most of the interpretations mentioned in this paper are related to combinatorial objects called lattice paths. What we mean by a *lattice path* is a sequence of contiguous unit steps of length n which traverse an integer lattice. The lattice paths are in the (x, y) plane such that all paths begin at the origin, $(0, 0)$, and never go below the *x-axis*. The length of each path is the number of unit-steps and the height corresponds to the y value of the point (x, y) at the end of the path. The symbols N, S, E and W denote unit-steps in the north, south, east and west directions, respectively. Thus, from the set of lattice paths, we have the following interpretation. The (n, k)th entry of C_0 (M and C, respectively) is the number of unit-step NS (NSE and NSEW, respectively) lattice paths of length n and height k.

Since C_0 is invertible, another infinite lower-triangular matrix R can be defined by $C_0 \cdot R = M$. The first few terms of R and $(C_0)^{-1}$ are

$$R = \begin{bmatrix} 1 \\ 1 & 1 \\ 1 & 2 & 1 \\ 2 & 3 & 3 & 1 \\ 4 & 6 & 6 & 4 & 1 \\ 8 & 13 & 13 & 10 & 5 & 1 \\ \vdots & \vdots & \vdots & \vdots & \vdots & \vdots & \ddots \end{bmatrix}, \text{ and } (C_0)^{-1} = \begin{bmatrix} 1 \\ 0 & 1 \\ -1 & 0 & 1 \\ 0 & -2 & 0 & 1 \\ 1 & 0 & -3 & 0 & 1 \\ 0 & 3 & 0 & -4 & 0 & 1 \\ \vdots & \vdots & \vdots & \vdots & \vdots & \vdots & \ddots \end{bmatrix}.$$

R can also be defined by $R = (C_0)^{-1} \cdot P \cdot C_0$ since $M = P \cdot C_0$. These relations and other matrix relations involving R are defined in section 2. A surprising fact about R is that the entries in the left most column, sequence $\{1, 1, 1, 2, 4, 8, 17, \ldots\}$, count ribonucleic acid (RNA) secondary structures of length n. As a result of the sequence, we summarize some of the combinatorial aspects of RNA secondary structures in section 3. Then, a lattice path interpretation of R is defined in section 4. Given the interpretation, a one-to-one correspondence between RNA secondary structures and lattice paths is constructed in section 5. Then, the first moments of R are computed and shown to be equal to the alternating Fibonacci numbers in section 6. We then conclude with applications of the Narayana numbers and noncrossing partitions in section 7.

2. Matrix Relations

Relations involving all of the above triangles are defined in this section. An outline of a proof of the relations is mentioned. The invertibility of C_0 is also mentioned.

Multiplying by R, the matrix relations $M \cdot R = C$ and $C \cdot R = H$ are defined. In the latter relation, H is another infinite lower-triangular matrix where the first few terms are

$$H = \begin{bmatrix} 1 & & & & & \\ 3 & 1 & & & & \\ 10 & 6 & 1 & & & \\ 36 & 29 & 9 & 1 & & \\ 137 & 132 & 57 & 12 & 1 & \\ \vdots & \vdots & \vdots & \vdots & \vdots & \ddots \end{bmatrix}.$$

The entries in the left most column, sequence $\{1, 3, 10, 36, 137, 543, \ldots\}$, count edge rooted polyhexes with n hexagons [9], [14]. These are graphs which are constructed by connecting n hexagons with certain restrictions. However, to remain within the context of lattice paths, we define the (n, k)th entry of H as the number of unit-step NSEWF lattice paths of length n and height k. The F denotes a forward unit-step. These lattice paths are 3-dimensional and they correspond to integer points (x, y, z) such that all paths begin at the origin, $(0, 0, 0)$, and never go below the (x, y) plane. The F steps are along the y-*axis*, and the height corresponds to the z value of (x, y, z) at the end of the path.

From all lattice path interpretations mentioned above and the associated matrix relations, we observe that right multiplication by R takes NS paths to NSE paths, NSE paths to NSEW paths, and NSEW paths to NSEWF paths. These path relations are illustrated as $C_0 \to M \to C \to H$ where the arrow means "goes to." In a combinatorial sense, R acts as a matrix transformation which transforms a selected set of unit-step lattice paths of length n and height k from 1-dimension to 2-dimensions to 3-dimensions. Likewise by left multiplication by P, the same transformation emerges since $P \cdot C_0 = M$, $P \cdot M = C$, and $P \cdot C = H$. The following proposition arises as a result of the matrix relations.

2.1. Proposition. *Given infinite lower-triangular matrices C_0, C, M, P and H, the following matrix relations are satisfied:*

$$\begin{array}{llcl} (a) & (C_0)^{-1} \cdot P \cdot C_0 & = & R \\ (b) & C_0 \cdot R = M & = & P \cdot C_0 \\ (c) & M \cdot R = C & = & P \cdot M \\ (d) & C \cdot R = H & = & P \cdot C \end{array}$$

Since the matrices are infinite it may not be obvious that the proposition is true. One way to prove the proposition is to consider that all of the matrices mentioned above are Riordan. *Riordan matrices* are infinite lower-triangular matrices made up of columns of the form $g(x) \cdot [f(x)]^i$ where $g(x) = 1 + g_1 x + g_2 x^2 + g_3 x^3 + \cdots$ corresponds to the left most column. For $i > 0$, the expression $g(x) \cdot [f(x)]^i$ corresponds to the *ith* column where $f(x) = 1x + f_2 x^2 + f_3 x^3 + f_4 x^4 + \cdots$. These columns are characterized as column generating functions (GFs) with integer coefficients f_i and g_i, and 1's along the main diagonal. A useful property of Riordan matrices is that they are invertible. Thus, the existence of $(C_0)^{-1}$ follows since C_0 is Riordan. Applying the Riordan matrix enumeration technique defined by Shapiro [18], or the Riordan array enumeration technique defined by Sprugnoli [22], the proposition can easily be proved. Likewise, using the same techniques, we can prove that the column GF associated with the left most column of H equals the GF associated with edge rooted polyhexes with n hexagons.

In the next two sections, the notions that motivate the correspondence between lattice paths and RNA secondary structures are mentioned. The relevant combinatorial aspects of RNA secondary structures are summarized, and then a combinatorial interpretation of R is given in terms of lattice paths.

3. RNA Secondary Structure

The single-stranded RNA molecule consists of a chain of base pairs derived from one of four bases (nucleotides): A (adenine), C (cytosine), G (guanine), and U (uracil) where A bonds with U, and G bonds with C. The linear sequence of such bases along the chain is defined as the *primary structure*. When an RNA molecule folds back on itself and forms new hydrogen bonds which form helical regions, the sequence is referred to as the secondary structure. As an example of secondary structure, we give the following RNA sequence s denoted as

$$s = CAGCAUCACAUCCGCGGGGUAAACGCU.$$

This sequence is referred to as a cloverleaf, in the biological literature, and is the secondary structure assumed by transfer RNA molecules. Two representations of s appear below in Fig. 1. In the figure we ignore the C, A, G and U and focus on the secondary structure.

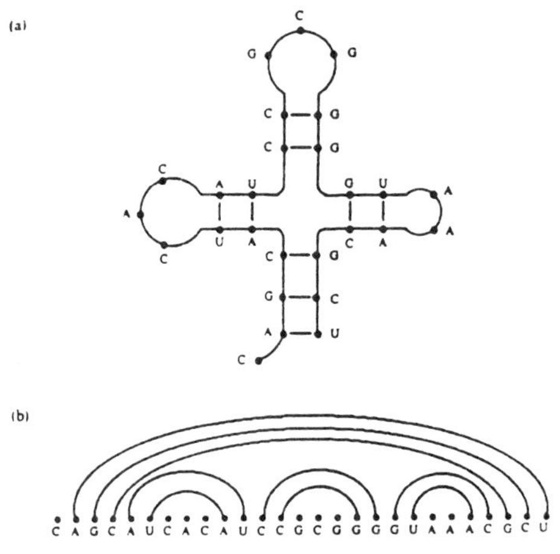

Fig. 1. Two representations of secondary structure.

In Fig. 1(a) the base pairs are indicated by dashes. In Fig. 1(b), the primary structure is given along the horizontal axis and the base pairs are shown as arcs. The above example, description and figure comes from Schmitt and Waterman [16]. The enumeration of secondary structures was studied, from a graph theoretic point of view, by Waterman [25]. He gives a graph theoretic definition of *secondary structure* as a planar graph defined on a set of n labeled points $\{1, 2, \ldots, n\}$. Thus, if $s(n)$ denotes the total number of secondary structures defined on n labeled points, then the associated recurrence relation is

$$s(n+1) = s(n) + \sum_{j=1}^{n-1} s(j-1)s(n-j) \tag{1}$$

for $n \geq 2$, and $s(0) = s(1) = s(2) = 1$. The first few values of $s(n)$ for $n = 0, 1, \ldots, 6$ are $1, 1, 1, 2, 4, 8$, and 17. Donaghey [3] notes that these numbers can also be computed by the following sum

$$\sum_{k \geq 1} \frac{1}{(n-k)} \binom{n-k}{k} \binom{n-k}{k-1}.$$

We call these numbers the RNA numbers. It turns out that these numbers are the same numbers as the entries in the left most column of R. The generating function derived from the recurrence relation is

$$s(x) = \frac{(1 - x + x^2) - \sqrt{1 + x + x^2)(1 - 3x + x^2)}}{2x^2} \tag{2}$$

Proofs of the recurrence relation, and generating function can be found in [10], [14] and [25].

4. Lattice Path Interpretation

Recursions for R are defined in this section by using the rule of formation of R. What we mean by rule of formation is a recursion which defines the way the entries of R are formed or computed. From the recursions, a combinatorial argument is proved showing that the entries of R denote the number of unit-step NSE* lattice paths of length n and height k. These NSE* lattice paths are explicitly defined later in this section. Then, the GF associated with the left most column of R is derived and shown to be equal to $s(x)$.

As examples of the way the elements of R are formed, we observe that the second column entry **6** is computed from $2+3+1$, and the left most column entry **8** is computed from $4+3+1$. The following illustrations:

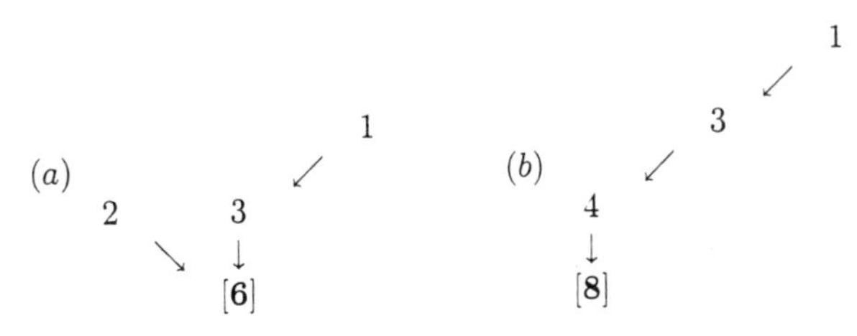

depict the rule of formation of the entries shown in the above examples. These patterns continue to form all of R. In general, the (n,k)th entry of R is formed or computed recursively by the following recursions. For $n \geq 0$ and $k \geq 1$

$$r(n+1,k) = r(n,k-1) + r(n,k) + r(n-1,k+1) + \cdots, \text{ and} \tag{3}$$

$$(4) \qquad r(n+1,0) = r(n,0) + r(n-1,1) + r(n-2,2) + \cdots$$

where $r(0,0) = 1$, and $r(n+1,k) = 0$ if $k > n+1$. Recursion 4 is defined for the left most column of R, and the associated column GF is denoted as $r(x)$. Recursion 3 is defined for the rest of the columns of R, and the associated kth column GF is denoted as $r(x) \cdot [f(x)]^k$ for $k > 0$. These associated GFs follow since R is a Riordan matrix. The GFs $r(x)$ and $f(x)$ are derived later in this section.

The NSE* lattice paths are now defined. These paths are also 2-dimensional and they have the same path restrictions as the NSE paths, mentioned in section 1, except for the additional restriction that consecutive N and S steps are **not** allowed. That is, a NSE* lattice path is a unit-step NSE lattice path which does not have any consecutive pair of NS steps. We combinatorially interpret the recursions in terms of these paths by letting $r(n,k)$ denote the number of unit-step NSE* lattice paths of length n and height k. Given the recursions and the interpretation, we can prove the following proposition.

4.1. Proposition. *For* $r(0,0) = 1$, $n \geq 0$ *and* $k \geq 1$, $r(n+1,k)$ *satisfy the following equations*:

$$(a)\ r(n+1,k) = \begin{cases} r(n,k-1) + \sum_{j\geq 0} r(n-j,k+j) \\ 0, \text{ if } k > n+1 \end{cases}$$

$$(b)\ r(n+1,0) = \sum_{j\geq 0} r(n-j,j).$$

Proof. Suppose we have a unit-step NSE* lattice path of length n and height k. Then, to form a new path of length $(n+1)$ and height k consider the following cases. Case (i): if we have a path of length n and height $k-1$, then on the last step there is 1 choice for height $k-1$ (the N step). In this case, all paths whose last step is N are counted by $r(n,k-1)$. Case (ii): if we have a path of length n and height k, then on the last step there is also 1 choice for height k (the E step). In this case, all paths whose last step is E are counted by $r(n,k)$. Case (iii): if we have a path of length $(n-1)$ and height $k+1$, then the last possible sequence of steps for height $k+1$ is ES (east, south). In this particular case, all paths whose last sequence of steps is ES are counted by $r(n-1,k+1)$. Case (iv): if we continue and have a path of length $(n-j)$ and height $k+j$, then the last possible sequence of steps for height $k+j$ is ES^j (east, south j-times). These sequences occur since there are no NS steps. Here, all paths whose last sequence of steps is ES^j are counted by $r(n-j,k+1)$. Combining all of the cases give all possible ways of forming a new $(n+1)$st path of height k. Applying the addition principle, recursion (a) is proved. Recursion (b) is proved by similar reasoning.□

Thus, the combinatorial interpretation of R is proved. Also by similar reasoning, we can prove the lattice path interpretations defined above for H, C, M and C_0 [12].

Explicit GFs for $r(x)$ and $f(x)$ are now derived. Recall that R is Riordan, so each column is of the form $r(x) \cdot [f(x)]^k$. By the rule of formation of R, the kth column GF is defined as

$$r \cdot f^k = x\left(rf^{k-1} + rf^k + xrf^{k+1} + x^2rf^{k+2} + \cdots\right).$$

Solving for f, we find $f = x + xf + x^2f^2 + x^3f^3 + \cdots$. Since the sum is a geometric series, we obtain $f = xf^2 + x(1-x)f + x$. Now, solving f in terms of $f(x)$ and simplifying we obtain $f(x) = x \cdot s(x)$. Similarly, the left most column GF is defined as

$$r = 1 + x\left(r + xrf + x^2rf^2 + x^3rf^3 + \cdots\right).$$

Simplifying this equation, we find that r expressed in terms of $r(x)$ is the same GF as equation 2, i.e., $r(x) = s(x)$. Thus, we have explicit GF representations for $r(x)$ and $f(x)$.

The GF $r(x)$ is associated with the unit-step NSE* lattice paths of length n and height $k = 0$. As a result of $k = 0$, $r(x)$ corresponds to the paths that return to the *x-axis*. In general, as a consequence of deriving $r(x)$ and $f(x) = x \cdot r(x)$, we can define for $k \geq 0$ the GF $\frac{r(x)}{1-x\cdot r(x)}$ whose associated sequence is $\{1, 2, 4, 9, 21, 50, \ldots\}$. These numbers are the values of the row sums of R and they count the total number of unit-step NSE* lattice paths of length n and height k [12]. However, of all NSE* paths, we are only concerned with those paths that return to the *x-axis*.

5. RNA Correspondence with Lattice Paths

A one-to-one correspondence between NSE* lattice paths and RNA secondary structures is constructed in this section. Since $r(x)$ and $s(x)$ denote the same GF, the order of the set of unit-step NSE* lattice paths of length n and height $k = 0$ and the order of the set of RNA secondary structures of length n are the same. So, the *nth* terms are such that $|s(n)| = |r(n, 0)|$. We can now state the correspondence theorem and the corresponding constructive proof.

5.1. Theorem. *There is a one-to-one correspondence between the set of RNA secondary structures of length n and the set of unit-step NSE* lattice paths of length n and height $k = 0$.*

Proof. First, we show how the unit steps of a NSE* lattice path of length n and height $k = 0$ are assigned. To establish the required correspondence, let s be a secondary structure of length n. Then, for a given s list the RNA sequence as a sequence of integers increasing in order from left to right as a primary structure along a horizontal axis and denote base pairs (or bondings) as arcs. Now, consider whether an integer is paired or unpaired. If an integer (or base) k is unpaired label the *kth* integer as an E step. Otherwise, if a base pairing of integers occurs where an integer i represents a pairing with a larger integer j, then label the *ith* integer as an N step and the *jth* integer as an S step. Therefore, the correspondence is setup according to the rules $k \rightarrow E$, $i \rightarrow N$, and $j \rightarrow S$ where the arrow means "corresponds to." The definition of secondary structure ensures that the NSE* paths do not have any consecutive pair of NS steps since no two adjacent points (i.e., with labels i and $i + 1$) can be connected by an arc, and no two arcs may intersect. Also, the paths are of height zero since each N step is paired with an S step.

As an example of the correspondence consider the sequence $\varsigma = ACAGUU$ where A bonds with U and C bonds with G. The bondings are indicated by dashes in the following graph:

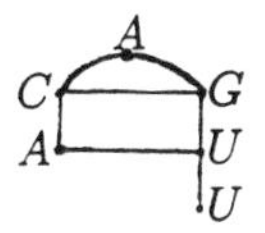

Now list ς as a sequence of integers along a horizontal axis as described above to obtain the following linear representation:

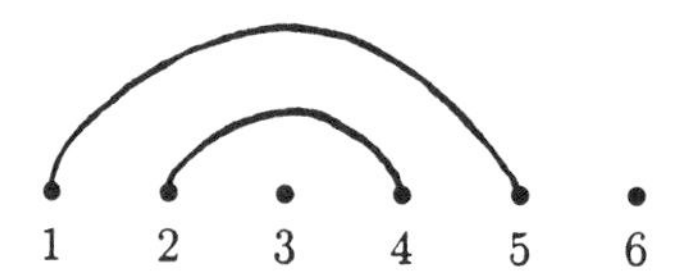

Applying the rules of the correspondence, the integers are assigned to the path steps as follows: 1 $\longrightarrow$N, 2 $\longrightarrow$N, 3 $\longrightarrow$E, 4 $\longrightarrow$S, 5 $\longrightarrow$S, and 6 $\longrightarrow$E. From the correspondence, we obtain the unit-step NSE* path NNESSE which has length 6 and height 0.

The correspondence is reversible. Thus, the correspondence is one-to-one and the theorem is proved.□

6. First Moments

The first moments (weighted row sums) of R are computed in this section. These moments can be used, in a combinatorial sense, to compute the average distance from the origin of all unit-step NSE* lattice paths of length n.

Multiplying R on the right side by the column vector $V^T = \{1, 2, 3, 4, 5, 6, 7, \ldots\}^T$, we make the following observation:

$$R \cdot V^T = \begin{array}{llllll} 1 \cdot 1 & & & & & = 1 \\ 1 \cdot 1 & +1 \cdot 2 & & & & = 3 \\ 1 \cdot 1 & +2 \cdot 2 & +1 \cdot 3 & & & = 8 \\ 2 \cdot 1 & +3 \cdot 2 & +3 \cdot 3 & +1 \cdot 4 & & = 21 \\ 4 \cdot 1 & +6 \cdot 2 & +6 \cdot 3 & +4 \cdot 4 & +1 \cdot 5 & = 55 \\ \vdots & \vdots & \vdots & \vdots & \vdots & \vdots \end{array}$$

Then, we conjecture that the first moments of R are defined by

$$R_n = F_{2n-1} = \sum_{k=0}^{n} k \cdot r(n-1, k-1) \text{ for } n \geq 1,$$

where F_{2n-1} denotes the alternating Fibonacci numbers 1, 3, 8, 21, 55, ... The conjecture can be proved using the Riordan matrix technique mentioned in section 2. The proof is outlined as follows. A Riordan matrix can be represented as a pair $[g(x), f(x)]$ where $g(x)$ and $f(x)$ are defined in section 2. A compositional functional $B(x)$ is obtained when a Riordan pair is multiplied on the right side by a GF denoted as $A(x)$. The GF $A(x)$ is associated with an appropriate column vector. Thus, $B(x)$ is defined as

$$\begin{aligned} B(x) &= [g(x), f(x)] * A(x) \\ &= g(x) \cdot A(f(x)) \end{aligned}$$

where the symbol ' * ' denotes Riordan matrix multiplication. The Riordan pair associated with R is the pair $[r(x), x \cdot r(x)]$ where $r(x)$ is the GF defined by equation 2 of section

3. The GF associated with the column vector V^T is defined by $v(x) = (1-x)^{-2}$. Then by Riordan multiplication, a compositional functional $F(x)$ is obtained and defined as

$$\begin{aligned} F(x) &= [r(x), x \cdot r(x)] * v(x) \\ &= r(x) \cdot v(x \cdot r(x)) \\ &= \tfrac{1}{1-3x+x^2}. \end{aligned}$$

Therefore, $F(x) = \frac{1}{1-3x+x^2}$ which is the GF for the alternating Fibonacci numbers [1], [5]. This proves the conjecture.

7. Other Applications

In the previous sections, we showed that the R triangle has matrix properties that are of combinatorial interest, and an application related to the RNA numbers. In this section, we mention other appearances of the RNA numbers related to noncrossing partitions, and the Narayana numbers.

The Narayana numbers are also of combinatorial interest and are defined as $N(n,k) = \frac{1}{n}\binom{n}{k}\binom{n}{k-1}$ for $n \geq 1$ and $k \geq 1$. These numbers can be put into infinite lower-triangular matrix form, denoted as N. The triangle N is not Riordan, and the first few terms are

$$N = \begin{bmatrix} 1 & & & & & \\ 1 & 1 & & & & \\ 1 & 3 & 1 & & & \\ 1 & 6 & 6 & 1 & & \\ 1 & 10 & 20 & 10 & 1 & \\ \vdots & \vdots & \vdots & \vdots & \vdots & \ddots \end{bmatrix}.$$

It is known that the row sums of N are the nth Catalan numbers C_n [11]. A combinatorial interpretation of N is that $N(n,k)$ counts the number of Dyck paths of length $2n$ with k peaks. A *Dyck* path is a path in the first quadrant, which begins at the origin, ends at $(2n.0)$, and consists of north-east and south-east steps. We note that the diagonal slices $1.1.1+1, 1+3, 1+6+1, \ldots$ of the N triangle give the first few RNA numbers. This can be proved by using generating functions, where the bivariate GF associated with the Narayana numbers is noted by Stanley [23].

To find a combinatorial relation between the RNA numbers and noncrossing partitions consider the set $[n] := \{1, 2, \ldots, n\}$. A partition π of $[n]$ is said to be *noncrossing* if $1 \leq a < b < c < d \leq n$ and if B_1 and B_2 are blocks of π such that $a, c \in B_1$ and $b, d \in B_2$, then $B_1 = B_2$. That is, given that the conditions are satisfied, a, b, c and d are all in the same block. As an example of a noncrossing partition of $[6] = \{1, 2, 3, 4, 5, 6\}$ consider $\pi = 1\ 5\ /\ 2\ 4\ /\ 3\ /\ 6$ where the slashes separate the blocks. The linear representation of π is illustrated above in section 5, where successive elements in the same block are joined by arcs.

Following Simion and Ullman [20], a word w of length $n-1$ over the alphabet $\{b, e, l, r\}$ can be associated with a noncrossing partition π. See the reference for detailed definitions of each letter in the alphabet. By eliminating the letter r and the consecutive pair of letters b and e from any potential word, another word w^* can be defined over the alphabet $\{b, e, l, \}$. The word w^* can also be associated with a noncrossing partition π.

For example, noncrossing partition $\pi = 1\ 5\ /\ 2\ 4\ /\ 3\ /\ 6$ is associated with the word $w^*(\pi) = bbleel$. A one-to-one correspondence can be constructed between the subset of noncrossing partitions associated with words w^* and the set of NSE* lattice paths. The correspondence is setup according to the following rules where $b \to N$, $e \to S$, and $l \to E$. Thus, via NSE* lattice paths, a relation is established between noncrossing partitions and the RNA numbers.

Several appearances of the RNA numbers related to planar trees are in [3]. Other combinatorial applications and interpretations of the RNA numbers are discussed in [2], [12] and [24].

Acknowledgment

This paper was developed during research on the author's dissertation. The author wishes to thank his advisor, Professor Louis W. Shapiro, and the combinatorics research group of Howard University for the many useful discussions pertaining to the material presented here.

References

[1] E.F. Beckenbach, *Applied Combinatorial Mathematics*, Wiley, New York, 1964.

[2] M. Bernstein and N.J.A. Sloane, *Some canonical sequences of integers*, Linear Algebra and its Applications **226-228** (1995) 57-72.

[3] R. Donaghey, *Automorphisms on Catalan trees and bracketings*, J. Combinatorial Theory, Series B **29** (1980) 75-90.

[4] R. Donaghey and L. W. Shapiro, *Motzkin numbers*, J. Combinatorial Theory, Series A **23** (1977) 291-301.

[5] R.L. Graham, D.E. Knuth and O. Patashnik, *Concrete Mathematics*, Addison-Wesley, Reading, MA, 1989.

[6] R.K. Guy, *Catwalks, sandsteps & Pascal pyramids*, preprint.

[7] R.K. Guy, C. Krattenthaler and B.E. Sagan, *Lattice paths, reflections, & dimension-changing bijections*, ARS Combinatoria **34** (1992) 3-15.

[8] P. Hanlon, *Counting interval graphs*, Trans. Am. Math. Soc. **272** (1982) 383-426.

[9] F. Harary and R. Read, *The enumeration of tree-like polyhexes*, Proc. Edinburgh Math. Soc. **17** (1972) 1-13.

[10] J.A. Howell, T.F. Smith and M.S. Waterman, *Computation of generating functions for biological molecules*, SIAM J. Appl. Math. **39** (1980) 119-133.

[11] T.V. Narayana, *Lattice Path Combinatorics with Statistical Applications*, University of Toronto Press, 1979.

[12] A. Nkwanta, *Lattice paths, generating functions, and the Riordan group*, Ph.D. Dissertation, Howard University, in progress, 1996.

[13] P. Peart and L. Woodson, *Triple factorization of some Riordan matrices*, Fibonacci Quart. **31** (1993) 121-128.

[14] F.S. Roberts, *Applied Combinatorics*, Prentice Hall, New Jersey, 1984.

[15] B. Sands, *Problem 1517**, Crux Mathematicorum **17** (1991) 119-122.

[16] W.R. Schmitt and M.S. Waterman, *Linear trees and RNA secondary structures*, Discrete Appl. Math. **51** (1994) 317-323.

[17] L.W. Shapiro, *A Catalan triangle*, Discrete Math. **14** (1976) 83-90.

[18] L.W. Shapiro, S. Getu, W.J. Woan and L. Woodson, *The Riordan Group*, Discrete Appl. Math. **34** (1991) 229-239.

[19] L.W. Shapiro, W.J. Woan and S. Getu, *Runs, slides and moments*, SIAM J. Alg. Disc. Meth. **4** (1983) 459-466.

[20] R. Simion and D. Ullman, *On the structure of the lattice of noncrossing partitions*, Discrete Math. **98** (1991) 193-206.

[21] N.J.A. Sloane and S. Plouffe, *The Encyclopedia of Integer Sequences*, Academic Press, San Diego, 1995.

[22] R. Sprugnoli, *Riordan arrays and combinatorial sums*, Discrete Math. **132** (1994) 267-290.

[23] R. Stanley, *Enumerative combinatorics*, Vol. II, Cambridge University Press, preliminary version, 1996.

[24] P.R. Stein and M.S. Waterman, *On some new sequences generalizing the Catalan and Motzkin numbers*, Discrete Math. **26** (1979) 261-272.

[25] M.S. Waterman, *Secondary structure of single-stranded nucleic acids*, Adv. Math. **I** (suppl.) (1978) 167-212.

Department of Mathematics, Howard University, Washington, D.C. 20059
E-mail address: nkwanta@scs.howard.edu

DIMACS Series in Discrete Mathematics
and Theoretical Computer Science
Volume **34**, 1997

Nupri as a Concurrent Interactive Theorem Prover*

Roderick Moten
Colgate University

Abstract

In this paper, we describe how to redesign the Nuprl Proof Developement System in order to perform concurrent inference. We focus on the design of the inference engine of Nuprl, the refiner. Our design consists of making the refiner a multithreaded program in which each thread behaves as a refiner. With the multithreaded refiner, we are able to develop tactics that execute concurrently. Tactics are programs employed by users to construct proofs within Nuprl. By having them execute concurrently, Nuprl will be the first interactive theorem prover with concurrent inference.

1 Introduction

Mechanical threorem proving employs mathematical logic in computer programs in order to prove theorems. With a mechanical theorem prover, mathematicians can prove theorems automatically or interactively. By automatically we mean that the theorem prover determines the existences of a proof of a conjecture without any human intervention. A user supplies the conjecture to the prover, and it replies whether the conjecuture is provable. In addition to the conjecture, the user may supply axioms and inference rules for the prover to use in determining the existence of a proof. An *interactive theorem prover* requires that the user supply methods of reasoning that the prover uses to construct a proof. Usually interactive theorem provers require that all reasoning be made using the proof rules of the object logic. The prover could be strict and require that the user supply only inference rules. This kind of interactive theorem prover is called a *proof checker*. Other interactive theorem provers allow the user to supply programs that generate inference rules that the prover carries out to construct a proof. Supplying programs to construct proofs in an interactive theorem prover was pioneered by the LCF theorem proving system [13]. In LCF, these programs were called *tactics*.

1991 *Mathematics Subject Classification*. Primary 68T15; Secondary 03B35.

*Support for this research was provided by the Office of Naval Research through grant N00014-92-J-1764.

Nuprl is an interactive theorem prover that uses tactics to provide automated deduction [7]. Nuprl was created to implement a constructive type theory similiar to Martin-Löf's type theory [17] to exploit the notion that proofs in constructive type theory can be converted into programs in a functional programming language [6, 11]. Some example mathematics that has been developed using Nuprl are abstract algebra [12], Ramsey's theorem [5], Higman's lemma [16], automata theory, and real analysis [9].

Using tactics in Nuprl allows one to develop formal proofs using intuitive methods. A single tactic can represent multiple steps of inferences and can use heuristics to select appropriate inferences. In Nuprl, users develop tactics using ML, a full featured functional programming language and *tacticals* [13]. Tacticals are tactic combinators; they allow users to create new tactics from existing tactics.

In this paper, we show how Nuprl can be modified to support *concurrent tactics*. A concurrent tactic is a single tactic constructed from several tactics that execute concurrently. To support concurrent tactics we developed a multiprocessor version of the Nuprl refiner, the inference engine of Nuprl, called the *MP refiner*.

2 Interactive Theorem Proving In Nuprl

In Nuprl, conjectures are represented as sequents, which we call *goals*. A goal is made up of a list of hypotheses and a term representing the conclusion.

$$h_1 \dots h_n \vdash c$$

Intuitively, the above goal means prove c true by assuming $h_1, \dots, h_n$ are true. A user can only supply goals to Nuprl that have no hypotheses. After the user enters a goal, Nuprl prompts the user to supply a tactic to generate a proof of the goal. When the user supplies the tactic, Nuprl executes the tactic. The tactic will either fail to construct a proof or construct a proof that has zero or more assumptions that make the proof true. Each assumption is a goal which we call *subgoals*. If there are subgoals, then Nuprl prompts the user to prove each of the subgoals. This continues until there are no subgoals to prove. To demonstrate proof construction in Nuprl, we will simulate part of the proof of the elementary number theory theorem for all $i \geq 8$ there exists $m, n \in \mathcal{N}$ such that $i = 3m + 5n$. We call this theorem *Stamps*. After we supply Stamps as a goal to Nuprl, Nuprl prompts us to supply a tactic by displaying the word BY.

```
>> ∀i:8... ∃m:N. ∃n:N. i = 3*m + 5*n.
   BY
```

We will prove the theorem by induction. Therefore, we apply the tactic *Induction "i"* to develop an induction proof over i. Appling the tactic generates two subgoals, the base case and the inductive case.

```
>> ∀i:8... ∃m:N. ∃n:N. i = 3*m + 5*n
    BY it Induction ''i''
       1. i: Z
       2. 0 < i
       3. 8 = i
   >> ∃ m:N. ∃n:N. i = 3*m + 5*n
   1. i: Z
   2. 8 < i
   3. m : N
   4. n : N
   5. i -1 = 3*m + 5*n
   >> ∃ m:N. ∃n:N. i = 3*m + 5*n
```

Proving these two subgoals will lead to an inductive proof of the initial goal. We can develop the proof in six steps using tactics. These six steps construct a formal proof of 605 primitive inferences from the Nuprl.

2.1 Tactics

Tactics are user defined programs written in ML, an acronym for Meta Language [13]. ML is a full featured functional programming language. In addition, Nuprl provides tactic combinators, called *tacticals*, which allows users to develop new tactics from existing tactics. Tacticals originated from LCF and are employed in various interactive theorem provers [10, 8].

Two common tacticals are THENL and ORELSE. THENL combines a tactic t and a list of tactics $[t_1; \ldots; t_n]$ into a single tactic, t THENL $[t_1; \ldots; t_n]$. Applying t THENL $[t_1; \ldots; t_n]$ to a goal g will first apply t to g to produce $g_1, \ldots, g_n$. Then t_i would be applied to the g_i for all i. ORELSE combines two tactics t and t' into a single tactic t ORELSE t'. Applying t ORELSE t' to g would apply t to g first. If the application failed to produce a proof then t' would be applied to g.

Although using tactics simplifies the development of formal proofs, tactics can be quite expensive. Some of the tactics provided in the hardware verfication suite developed by Leeser and Aagard [1] to provide an environment for constructing and verifying circuit designs in Nuprl can take up to 20 minutes to execute. We use parallel techniques to possibly improve the running times of tactics.

3 Concurrent Tactics

A *concurrent tactic* is a single tactic made of several tactics that execute concurrently. Concurrent tactics can be constructed using two new tacticals, PTHENL and PORELSE. Both PTHENL and PORELSE have the same behavior as THENL and ORELSE, but PTHENL and PORELSE execute their arguments in parallel. In other words, t THENL $[t_1, \ldots, t_n]$ and t PTHENL $[t_1, \ldots, t_n]$ generate the same proof, and t ORELSE t' and t PORELSE t' generate the same proof. However, executing t PTHENL $[t_1, \ldots, t_n]$ on a goal g will execute each of the t_i's in parallel. Likewise, t PORELSE t' executes t and t' in parallel. To support concurrent tactics,

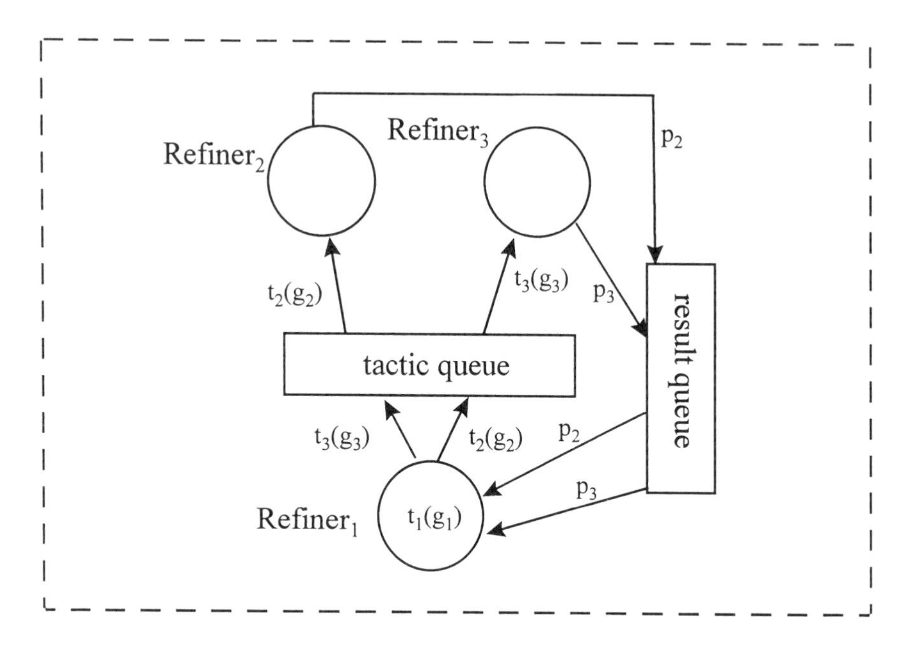

Figure 1: Model of MP Refiner

we implement the MP refiner, a multiprocessor version of the refiner, Nuprl's inference engine.

3.1 Implementation Model of the MP Refiner

The MP refiner is a multithreaded program. The diagram in Figure 1 depicts the implemenation model of the MP refiner. The MP refiner in Figure 1 has three threads behaving as sequential refiners. In general, the MP refiner can contain at least two threads. A refiner thread spawns a tactic application to be executed by another refiner thread by placing the tactic and goal in the *tactic queue.* In Figure 1, Refiner$_1$ spawns the applications of t_2 and t_3 to g_2 and g_3, respectively. When a refiner thread is idle, it will deque a tactic with its corresponding goal from the tactic queue and perform the application. The results of the application (p_2 and p_3) are placed in the *result queue.* Meanwhile, the refiner thread that spawned the tactic (Refiner$_1$) can do other work like carry out another tactic application ($t_1(g_1)$). At some point the refiner thread that spawned the tactic application retrieves the results of the spawned tactic application from the result queue.

To write concurrent tactics, we extend ML to contain primitives for spawning tactics to be executed on other refiner threads. A ML program can use these primitives to request that a tactic be executed by a different refiner thread. We use these primitives to implement two new tacticals `PTHENL` and `PORELSE`.

3.2 Implementing the MP Refiner

The MP refiner will be implemented using Standard ML of New Jersey [4]. We modified the runtime system of Standard ML of NJ to support Solaris threads [18] following Morrisett's and Tolmach's generic MP interface [14]. The modification allows one to create multithreaded Standard ML programs that can execute in parallel. The implementation of the MP refiner follows the mathematical description of the refiner [15]. The mathematical description is a specification of the refiner and is based on the reflected Nuprl type theory [3, 2]. An intent of the mathematical description of the refiner is to give people the ability to understand the inference strategy of Nuprl without having to use it.

The MP Refiner will be part of Nuprl 5. Nuprl 5 is an initiative to give Nuprl an open distributed achitecture. It makes the components of Nuprl, the editor, the library, and the refiner, to be interchangeable. In other words, one could take different components of Nuprl created by different vendors and link them to create a single proof system.

4 Conclusion

The MP refiner will make Nuprl the first interactive theorem prover with concurrent tactics. For our preliminary tests, we developed concurrent versions of two sequential tactics that have average running times of 0.5 seconds and 7.8 seconds. The concurrent versions of these tactics have average running times of 0.9 seconds and 6.2 seconds, respectively. In the future, we intend to conduct thorough experiments based on cost models of tactics. For each cost model, we will obtain a sample of tactics and compare the execution times of the sequential and concurrent tactics. In addition, we intend to investigate the effect of concurrency on soundness and consistency.

References

[1] M. Aagaard, M. Leeser, and P. Windley. Toward a super duper hardware tactic. In *Higher Order Theorem Proving and its Applications.* Lecture Notes in Computer Science, 780, Springer-Verlag, 1993.

[2] S. Allen, R. Constable, and D. Howe. Reflecting the open–ended computation system of constructive type theory. In F. Bauer, editor, *Logic, Algebra and Computation*, pages 267–288. Springer-Verlag, 1991.

[3] S. Allen, R. Constable, D. Howe, and W. Aitken. The semantics of reflected proof. In *Fifth Annual IEEE Symposium on Logic in Computer Science*, pages 95–105. IEEE Computer Society Press, 1990.

[4] A. W. Appel and D. B. MacQueen. Standard ML of New Jersey. In *Programming Language Implementation and Logic Programming: 3rd International Symposium*, pages 1–13. Springer-Verlag, 1991.

[5] D. A. Basin. *Building Problem Solving Environments in Constructive Type Theory.* PhD thesis, Cornell University, 1989.

[6] J. Bates and R. Constable. Proofs as programs. *ACM Transactions on Programming Languages and Systems*, 7:113–136, 1985.

[7] R. L. Constable et al. *Implementing Mathematics with the Nuprl Proof Development System.* Prentice Hall, Englewood Cliffs, NJ, 1986.

[8] C. Cornes et al. The Coq Proof Assistant Reference Manual: Version 5.10. Unpublished, 1995.

[9] M. B. Forester. Formalizing constructive real analysis. Technical Report TR93-1382, Cornell Computer Science Department, 1993.

[10] M. Gordon. Hol: A proof generating system for higher-order logic. In *VLSI Specification, Verification and Synthesis*, pages 73–128. Kluwer Academic Publishers, 1988.

[11] W. Howard. The formulae-as-types notion of construction. In J. Seldin and J. Hindley, editors, *To H.B. Curry: Essays on Combinatory Logic, Lambda Calculus and Formalism*, pages 479–490. Academic Press, 1980.

[12] P. Jackson. *Enhancing the Nuprl Proof Development System and Applying it to Computational Abstract Algebra.* PhD thesis, Cornell University, 1995.

[13] A. J. Milner et al. *Edinburgh LCF.* Lecture Notes in Computer Science 78. Springer-Verlag, New York, 1979.

[14] J. G. Morrisett and A. Tolmach. A portable multiprocessor interface for Standard ML of New Jersey. Technical Report CMU-CS-92-155, School of Computer Science Carnegie Mellon University, 1992.

[15] R. Moten. *Concurrent Refinement in Nuprl.* PhD thesis, Cornell University, 1997.

[16] C. Murthy. A constructive proof of higman's lemma. Technical Report TR89-1049, Cornell University, 1989.

[17] Per Martin-Löf. Constructive mathemetics and computer programming. In *Logic, Methodolgy, and Philosophy of Science VI*, pages 153–175. North-Holland, 1982.

[18] M. L. Powell et al. Sunos multi-thread architecture. In B. J. Catanzaro, editor, *The Sparc Technical Papers*, pages 339–372. Springer-Verlag, 1991.

Roderick Moten is on leave of absence as a Ph.D student in the Computer Science Department at Cornell University while he finishes a one year appointment as a visiting instructor in the Department of Computer Science at Colgate University.

Part III

Historical Articles

DIMACS Series in Discrete Mathematics
and Theoretical Computer Science
Volume **34**, 1997

Yesterday, Today and Tomorrow†

LEE LORCH

ABSTRACT This is a discussion of the domestic and international contexts within which the African American mathematical community has struggled and is struggling. The effect of general social forces on scientific opportunities is emphasized. Stress is laid upon the importance of supporting the civil rights movement so as to protect, hopefully to expand, minority opportunities in mathematics. These are the mass movements which have forced open some windows of opportunity, windows powerful forces are now seeking to close.

Dear Friends,

I am very honoured to be here. I am very honoured to have been asked to speak this evening, another heartwarming sign of the friendship that I have enjoyed with the African American community these many years. I am very honoured to be the successor to J. Ernest Wilkins Jr. who was the keynote speaker at the inaugural conference held last year in Berkeley.

He is unable to be here this evening, as are a number of others who would also like to be with us. Like myself, he is in the fourth phase of life. First comes childhood, then youth, then middle age and finally "My, you're looking well!" J. Ernest is indeed looking well. Unfortunately, at this time his wife is not well and so he is unable to be with us.

His accomplishments are inspiring, his career illuminating. With a Ph.D. in mathematics from the University of Chicago (1942), when not yet nineteen, no research university would offer him a job. The war on, he was able to get into industry; there were hardly any mathematicians in or available for industry then. Changing from pure mathematics, he became an applied mathematician and an engineer, becoming expert in both.

---

†1991 Mathematical Subject Classification. 01A67, 01A80.
Adapted from keynote address, June 27, 1996

After the war, he approached the University of Michigan which was anxious to build its applied mathematics program. Applied mathematics was in very short supply in the US until well after the Second World War. Before the war, there may not have been more than about half a dozen mathematicians in US industry. The war and post-war expectations called for a crash program. The Head of the Michigan Mathematics Department replied saying that he was very impressed with Dr. Wilkins's credentials but that "the times were not ripe" for appointing an African American! Here was an expert in applied mathematics with several years of industrial experience, a rarity in those days. But racism triumphed over scientific needs. Now Dr. Wilkins is a member of the élite National Academy of Engineering and a Professor at Clark Atlanta University.

To be fair, years of struggle, previous and contemporary, reinforced by the impact of the war and rapidly changing domestic and international conditions, had ripened the times a bit even then. Wade Ellis Sr. was appointed to Oberlin College, the first Black mathematician in our time to cross the line. And in the neighborhood of this Conference, Princeton University began then to enrol Black students whom it had long excluded, and eventually Black faculty.

This was quite a switch for Princeton. It is fairly well known that Paul Robeson, born and raised in Princeton, was not admitted to Princeton University. He went instead to Rutgers, there to shine brilliantly in scholarship and athletics, a preview of the magnificence of his life and character. Not so well known is that when David Blackwell went on a post-doc to the Institute of Advanced Study, then sharing facilities with Princeton University, the President of Princeton complained to the Institute of this "abuse of Princeton's hospitality" [9,p. 179]! Dr. Blackwell became the first Black in any field to be elected to the prestigious National Academy of Sciences and is now Professor Emeritus at the University of California-Berkeley. He served many years as Head of the Howard University Mathematics Department.

This Conference, like its predecessor, is inspiring. It is more diverse than the mainstream mathematics gatherings. We have young and old, junior and senior faculty,undergraduate and graduate students and we are all talking openly to one another, learning from one another, developing friendships, establishing badly needed networks, exchanging experiences, laying the basis for mentoring at all levels, not forgetting junior faculty seeking tenure, nor students needing to know what it's like out there and to have encouragement and support in taking the plunge. The gender distribution is more equitable, the "racial mix" more of a mix (albeit reversed), at these Conferences sharply focussed on the African American community than at supposedly "general" meetings, another proof of the need for these Conferences.

The Conferences have the support not only of those here but also of unwilling absentees. One of last year's leaders, Raymond Johnson, Chair of the Mathematics Department at the University of Maryland, had to go to a meeting (GEM) being held right now in Washington, D.C., to drum up financial support for Black graduate students. I expect some in this audience to be among the beneficiaries of his efforts.

Dr. Etta Zuber Falconer, Head of Spelman College's Science Division, winner last year of the prestigious Hay Award of the Association for Women in Mathematics, has to look after an elderly relative who requires her attention this week. But Dr. Falconer will be represented here to some extent. I'll be using data which she collected for a lecture she gave at the University of Wisconsin-Madison in May when she was awarded an honorary doctorate. That text is included in this volume [5]. I look forward to her being a keynote speaker at an early one of our successor conferences.

There is another absence, this very sad. Dr. Vivienne Malone Mayes died prematurely, age 63, just before the Berkeley Conference. Dr. Falconer and I wrote an obituary article which appeared in the *Newsletter of the Association for Women in Mathematics* [7]. The organ of the Mathematical Association of America, *Focus,* later reprinted it [8], as abbreviated by the editors. Some of the deletions contain concrete information about the discrimination she had to overcome and the methods necessary to do so. In a moment I'll cite some of the relevant deletions.

It wasn't simple to continue to a mathematics Ph.D. when Dr. Mayes obtained her Fisk M.A. in 1954. Where to find financial support? Where even to go where you'd be at least tolerated? You'd have to be both tough and talented. First she taught for some years at Paul Quinn College (operated by the African Methodist Episcopal Church) in her home town of Waco, Texas. Wanting to keep up in mathematics, but not yet seeking a degree, she sought to take some courses at Waco's Baylor University. Baylor said no, claiming that as a private institution it was not bound by hard-won Supreme Court decisions on public education. Eventually, she was able to get into the University of Texas, a public university bound by those court rulings. This was to be no bed of roses.

One of the deletions from [7] notes some of the thorns:

> In graduate school she was very much alone. In her first class, she was the only Black, the only woman. Her classmates ignored her completely, even terminating conversations if she came within earshot. She was denied a teaching assistantship although she was an experienced and excellent teacher. She wrote further: "I could not join my advisor and other classmates to discuss mathematics over coffee at Hilsberg's cafe.... Hilsberg's would not serve Blacks. Occasionally, I could get snatches of their conversations as they crossed our picket line outside the cafe." She "could not enrol in one professor's class. He did not teach Blacks."

Overlooking all this, one of her professors, complaining against the civil rights demonstrations, said to her: "If all those out there were like you,hard-working and studious, we wouldn't have any problems." Her reply: "If it hadn't been for those hell-raisers out there, you wouldn't even know me." [11]

Both [11] and [12] arise from invited talks she gave to the Association for Women in Mathematics. As [7] states, they "describe not only her own journey but that of the collective of Black women, indeed of all women and all Blacks. Written many years ago, they cast a penetrating light on the past and the present. They are must reading still today—and for everybody."[7]. Related materials can be found in [3,5,6,9] where further references are provided.

Her reply to the professor who didn't like the civil rights demonstrators goes to the heart of the matter. The opportunity to go to school, the opportunity to vote, the affirmative action programs, none of these originated within academia or the business world. They arose out of the sacrifice and mass movements in the minority community, allied with other progressive forces. They came about because many thousands had the courage to organize, to withstand fire hoses, attack dogs, tear gas, police batons—and to do it again and again and again.

What we have today didn't emerge from scholarly articles, but from struggles in which the entire Black community was sacrificially involved. If we are going to keep what has been obtained and to go forward to what is still desperately needed, we mustn't forget that. We must maintain and strengthen alliances within the community and between the community and all progressive forces. We must be sensitive to, and active concerning, the problems of poverty, unemployment, abuse, with which the community is struggling. A mountain peak needs a strong, solid base for its very existence.

Dr. Mayes wrote in 1975 [11] that things were going well for her at Baylor University where she had secured a faculty position in 1966, five years after being refused there as a student. By then the struggle had achieved a whole body of new legislation protective of minorities and women. And, as she explained, "An additional safeguard of my welfare has been yearly visits by representatives of the [federal] government. They have checked salaries and promotions to determine if I was being subjected to any discrimination."

That was 1975. Beginning around 1981 things began to change. Her situation worsened. Reagan had become President. He and his successor Bush wanted to repeal civil rights legislation, but too many Congressmen recognized the threat that would mean to their chances of reelection. A different strategy emerged: funding for the enforcement agencies was cut, enforcement lagged. Both vetoed civil rights bills enacted by the Congress. The message was clear.

The opponents of racism and sexism, the voices of the excluded and marginalized, the labor movement, the women's movement, the compassionate, all were now on the defensive.

This is not to suggest that we were back to where we had been before. Many important factors prevented that. It might be well to look at some of them, without forgetting the dangers still out there.

A dramatic case in point is the famous Supreme Court decision of May 17, 1954, in *Brown v. Board of Education.* Why did a unanimous court declare segregation in public education unconstitutional then when in the equally (in)famous 1896 case of *Plessy v. Ferguson* it had upheld segregation? Wiser or more compassionate judges? Better lawyers?

The social setting of these decisions provides more objective reasons. By the 1890s in the US, well after the liquidation of the reconstruction governments following the Hayes-Tilden Compromise of 1876, Blacks were totally powerless, without strong organizations,long since deprived well-nigh universally of the right to vote. Still primarily resident in the South, their voices had been stilled, their hopes smashed, their rightless poverty turned into others' wealth.

The international scene was just as bleak. Africa had been largely colonized, cynically divided among the European powers, its indigenous people without power in their own lands. The Indian sub-continent, home today to several large countries, was in the hands of the British Raj. The West Indies were the British West Indies. The US was about to displace Spain as the master of Cuba, the Platt Amendment giving it the "right" to intervene whenever it wished. Even China had been forced into humble subservience and required to yield extraterritorial rights to foreign powers. Great Britain waged successfully the Opium War to force China to allow the continuation of that vile traffic.

Neither within US borders, nor internationally, could be found any forces to compel respect and decent treatment for its African American or indeed Asian American or Hispanic or Native American population. Immigration laws also reflected this.

The position of African Americans as isolated, voteless, poverty-stricken, exploitable cheap labor, devoid of international support, seemed carved in stone in those days. US policy and its judiciary institutionalized this horror.

But little is cast in stone and the world can be made to change. True, between 1896, the year of *Plessy*, and 1954, the year of *Brown*, there had been no US legislation, no constitutional amendment adopted, which would have an iota of legal impact on the 1896 decision requiring its reconsideration. But a saying which has survived from that era, reminds us that "The Supreme Court follows the election returns."

By 1954 the situation had changed drastically from 1896, the result of vast struggles in both arenas. Many countries had freed themselves from colonial rule following the second world war. Migrations from the South to the North where it was possible to vote had given African Americans decisive voting power in states like Illinois, Indiana, Michigan, California, New York, Ohio, Pennsylvania which could determine Presidential elections and affect the composition of the federal Congress as well as state governments. Growing demands, accompanied by bitter and sacrificial struggle, by minorities for civil rights, their growing entrance into the labor force and, along with that, growing support from the labor movement, had all combined to make minorities a domestic political force of serious proportions.

The passion of the US government to extend and consolidate political and economic influence in the former colonies required Washington to take into account the sharp eyes that people of color around the world were casting on US domestic policy. The US authorities had to signal to the world and concede to its own struggling minorities some visible change.

The political winds had changed. They can change again. Now the US is the only superpower; its leaders may not feel much concern for world opinion. The Helms-Burton Act indicates that. In seeking to impose its legislation extra-territorially, the US has disregarded overwhelming votes in the UN General Assembly and the Organization of American States (the very first time that the OAS has condemned any US behavior) and loud protests from its closest allies. The mighty US government has put on its statute books punitive pressures upon the entire world in its passion to destroy the independence of little Cuba.

Domestically, well organized, unrelenting and massive onslaughts have been mounted on the rights of minorities and women, on the poor, on the labor movement. Affirmative action bears much of the brunt. Some academics, often with foundation support, busy themselves with well publicized deriding of the capacities of the victims, on the need (and obligation) which society at large has to nourish those capacities.

Various politicians and demagogues find, in the rapidly growing gap between rich and poor, enticing opportunities to advance themselves and enrich their sponsors. Confusing and poisoning the social atmosphere, they seek to direct the wrath of one set of victims away from the social order onto another set of victims. Divide and conquer.

Recent decisions of the US Supreme Court have reduced drastically the scope of affirmative action and have rendered other decisions affecting minorities adversely [14].

The Governor of California has cancelled over 150 affirmative action programs. The Regents of the University of California have withdrawn affirmative action programs for minorities and women. The "California Civil Rights Initiative," intended to kill all state affirmative action programs benefiting minorities and women was adopted on the November 1996 ballot.

The deterioration in the administrative and legal framework has a demoralizing as well as a palpable effect. Minority applications for admission to the University of California-Berkeley have declined. Quite a bit of useful information has been published in *Science*,organ of the American Association for the Advancement of Science (AAAS) [14, November 4,1994; October 6, 1995; March 28, 1996]. It can be retrieved also from the internet, as noted in the references. The internet hosts also the affirmative action brief of the American Civil Liberties Union (ACLU) [1].

I don't wish to comment on the growing threats in the US as if the province of Ontario, where I live, were immune. The new provincial government inaugurated its term by cutting welfare 22%, repealing equity (affirmative action) and anti-scab legislation and is about to gut rent control. We are all in the same boat.

The wide range of the attacks on the social net emphasizes both the need and the possibility of strengthening alliances with the broader community. The cutbacks in support for education, the increase in university tuition fees, to say nothing about the lack of adequate health care and job opportunities, all hit minorities and workers generally particularly hard.

This is an old and yet constantly worsening saga. One example from "yesterday": Elbert Cox was the first African American to attain the Ph.D. in mathematics (Cornell, 1925). Kelly Miller would have been much earlier. He was in the mathematics graduate program of Johns Hopkins University 1887–1889 but had to drop out when the tuition fee was doubled from $100 to $200, an amount equivalent to several thousand dollars today.

He became a high school mathematics teacher, then a professor (first of mathematics, later of sociology) at Howard University where he was also Dean. He had a satisfying and influential career. But poverty had barred his way to mathematical research. Countless others have had the way barred even to entertaining any dreams.

There are those—largely non-minority—who claim that underrepresented minorities no longer need special attention and are not entitled to compensatory assistance. In her Madison lecture [5], Dr. Falconer presented data culled from surveys made by the American Mathematical Society (AMS) and the like which, to put it mildly, do not support this view. Underrepresented minorities (African American, Hispanic, Native American) account all told for less than 3% of the mathematics doctorates awarded to US citizens in each of the years 1992-93,1993-94,1994-95, although they constitute nearly 27% of the population. Their presence in the work force is even more invisible. Of those holding mathematical doctorates,African Americans constituted 1.3%, Hispanics 1.5% and Native Americans 0.2% of those employed in 1991. Her data quantify the acceleration in the dropout rate as higher and higher educational levels are reached.

I have not mentioned so far the situation within the mathematics organizations. Some of that I discussed in a talk [10] at the 100th annual general meeting of the American Mathematical Society (AMS). Even there I alluded to the social setting, since it is in that framework that the opportunities for underrepresented minorities and women will be created or lost. There is one paragraph [10,p.87] that I would like to repeat here:

> The Canadian and US governments have apologised, as indeed they should, for the internment of their citizens of Japanese descent during World War II. I know of no plans to apologize for the centuries of slavery and discrimination inflicted upon those of African descent. Some day there will be an African country with the same economic and political clout as contemporary Japan. Then we can look forward to similar apologies to the descendants of Africa.

But our mathematical organizations could apologize for past behavior before then.

The bulk of [10] deals with the specifics of mathematics meetings. Its title emphasizes that those meetings have moved "toward," not "to," inclusiveness. Some of what had to be done to initiate movement is recounted there.

It is true, and praiseworthy, that both the AMS and the Mathematical Association of America (MAA) have now accepted obligations to encourage minorities and women. But it is also true that there has been no African American on the AMS Council for over ten years. Only four have served in the one hundred years of the AMS existence.

True too is it that the days are gone when AMS and MAA meetings were organized on the assumption that Blacks would not participate. Some details are in [10] and in [13, Appendix 2]. But it is also true that when the history of the MAA Southeastern region was published in 1992 [15], its author refused to use the documentation provided via Dr. Sylvia Bozeman of past racist practices. It would, he said, reopen old wounds.

The idea that the African American constituency would be hurt (to say nothing of the historical distortion) by having its grievances dismissed didn't seem to enter his thinking. His reaction is widespread. The Texas mathematician who, as Dr. Mayes recorded [11], would not teach Blacks has been celebrated in publications and film, but his well-known racism isn't mentioned and may be forgotten with the passage of time.

Thanks to the activities of our African American colleagues in Atlanta, a Supplement to the MAA history booklet has since been published [2]. It is available from the MAA office.

Small wonder that the community established many years ago the National Association of Mathematicians (NAM) to advance the Black presence in US mathematical life. A similar motivation brought into being the Association for Women in Mathematics (AWM) whose Hay Award Dr. Falconer won and on whose executive some African American women have served.

The AMS is particularly unrepresentative in its leadership and committee structure.It is only natural that program committees, e.g., would invite speakers from among those in the orbit of committee members. The first time that Dr. Wilkins was invited to address a joint meeting of the AMS and MAA was fifty years after his Ph.D. and many years after he had been elected to the National Academy of Engineering. It was Dr. Raymond Johnson's presence on the program committee for that meeting that led to this long overdue invitation.

It is in the colleges (junior and full term) and universities that professional hope or despair will be created. Access is of fundamental importance. Something must be done to end the social obscenity which finds more African American young men in the clutches of the prison system than in college, which is more interested in building prisons than schools. Further, even for those in a position to continue past secondary school, what atmosphere do they encounter, what support (financial and moral) is there, what jobs will they get? Will they get tenure or merely pass through a revolving door?

The majority institutions are now very tough. They deny good people tenure and replace them with other good people at beginning wages. This can also be a shield for racism and sexism. The schools can always have some token on hand, but in junior positions, lacking influence, perhaps even fearful of taking stands lest they jeopardize tenure chances. We have to help our junior faculty colleagues avoid this trap.

While on the subject of jobs, I would like to urge serious attention to the Historically Black Colleges and Universities (HBCUs). A significant number, at least 47, of mathematics Ph. Ds did their undergraduate work at HBCUs. These institutions provide a sense of community, nurture hope and build self-confidence.One, Howard University, now offers the Ph.D. in mathematics. At HBCUs there can be Black leadership in curriculum formation.

Of course, all minorities, African American and others, are entitled to be and should be everywhere. And it is true that the majority institutions are far better financed and can provide greater research opportunities than the typical HBCU. In the majority institutions I hope that minority faculty members, including mathematicians, will influence the general curriculum, not only their own departments.

It is important even for mathematics students to have the African American presence seen in literature, history, sociology, world events, everywhere, as well as in science. This would also offer a more accurate picture of the world and provide a better education for all, majority and minority alike.

Of course, even HBCUs are not free from the society in which they function. Funding and other pressures take their toll in many ways. In my time, many had all white Boards of Trustees (and intrusive ones at that) appointed by segregationist governors. White majorities on Boards were common. There is a big difference between attending or working at an institution and controlling it. Many stories could be told.

Dr. Dean has suggested that I say something about my dismissal from an HBCU in 1955, on the heels of the Supreme Court decision of May 17, 1954. Other opponents of segregation lost their jobs in other HBCUs in those days as well. In fact, the entire US academic and educational system was bereft of academic freedom and operated under Cold War and racist pressures.

Following the Supreme Court decision, my wife and I sought to enrol our daughter in the public school nearest the Fisk campus where we were living. A subpoena from the House Committee on Un-American Activities (HUAC) came swiftly, as did local newspaper outcries. Hysteria was generated, nominally attached to my refusal to tell HUAC whether I had ever been a member of the Communist Party. This brought an indictment for "Contempt of Congress," of which I was later acquitted following trial in Federal District Court.

One of the local whites on the Board threatened his public resignation from the Board, with consequent threats to agency funding, unless I was fired. Soon the white majority rallied solidly around him. There was great opposition to my dismissal, from the community, from the weekly newspapers, from colleagues, from prominent Black members of the Board, from Alumni. But the pressures were great.

Such episodes demonstrate anew the importance of the overall, apparently external, social setting to the internal functioning of HBCUs (and all other institutions) to their opportunity to develop mathematics, mathematics training and mathematical opportunities. John Donne's celebrated observation "No man is an island entire unto himself" applies to social structures as well.

Dr. Percy Julian, the distinguished chemist who became the second Black ever elected to the National Academy of Sciences, was among the Trustees who fought against my dismissal. Dr. W.E.B. DuBois, Fisk's most celebrated alumnus, was among the many alumni who came to my support. His statue now adorns the Fisk campus: In his home state, the University of Massachusetts has given his name to its library and published his collected works. He would be pleased to note in this audience the accomplished African American Head of that university's Mathematics Department,Dr. Donald St.Mary, especially since he is not alone in that Department.

The countenances of Dr. DuBois and Dr. Julian now grace US postage stamps in the Black Heritage series.

There is a campaign to commemorate Paul Robeson in this fashion. His intrinsic merits demand it; the US government, which persecuted him in his lifetime, owes it. Supporting letters can be sent to: Citizens Stamp Advisory Committee, US Postal Service, 475 L'Enfant Plaza, Room 4474E, Washington, D.C., 20260-2437.

Dr. DuBois had his own troubles with the government, plenty of them. One of his books [4] details how he was arrested, handcuffed and fingerprinted at age 83. He was Chair of the Peace Information Center headquartered in Paris. Peace was a dirty word in the US of the 1950s. The resulting felony charge against him was so specious that the judge dismissed the case without even letting it go to the jury. But the defense had cost, in addition to anguish and energy, a tidy sum of money.

As he wrote, "justice is not for the poor." In 1903 he had declared that "the problem of the Twentieth Century is the problem of the color-line." Expanding this a half century later, he embedded it, with no loss of emphasis, in the class structure of society. He sought its solution in the creation of a society free of exploitation and greed.

He was always optimistic, nor can we afford the luxury of pessimism. From the struggles of centuries there derives a rich inheritance and attendant obligations.

Easy times are not at hand. But we are not mere bystanders, nor are we without potential power. These Conferences have brought the African American mathematical research community together at all career levels. Its members know one another, can and must help, mentor and guide one another, befriend one another, and one another's students.

All this is necessary, but not sufficient. To this must be added concern for, understanding of, alliance with, the rest of the community and with all forces in society striving to create a better life. Benjamin Franklin put it succinctly, in his day literally: "If we do not hang together, we shall assuredly all hang separately."

I am confident that this generation will build on its heritage and display the heroic solidarity that the civil rights movement, which brought current opportunities, has bequeathed. Together we can shape tomorrow.

REFERENCES

1. American Civil Liberties Union, 132 West 43rd Street, New York, NY 10036; http://www.aclu.org/issues/racial/hmre/html.

2. S. Bozeman, E.Z. Falconer, A. Shabazz, J.E. Wilkins, Jr., *A History of Minority Participation in the Southeastern Section*, MAA, April 1995.

3. J.A. Donaldson, *Black Americans in Mathematics,* A Century of Mathematics in America, Part III, AMS, 1989, pp. 449–470.

4. W.E.B. DuBois, *In Battle for Peace*, 1952 (Kraus-Thomson Reprint, 1977).

5. E.Z. Falconer, *The Challenge of Diversity,* University of Wisconsin-Madison, May 1996, this volume.

6. ____, *Views of an African American Woman on Mathematical Meetings*, A Century of Mathematical Meetings, AMS,1996, pp. 79–82.

7. E.Z. Falconer and L. Lorch, *Vivienne Malone-Mayes in Memoriam*, AWM Newsletter, v. 25, no. 6 (Nov/Dec 1995), pp. 8–10

8. ___ and ___, *Vivienne Malone-Mayes Pathfinder*, Focus, v. 16, no. 3 (June 1996), p. 8, MAA.

9. P. Kenschaft, *Black Men and Women in Mathematical Research*, J. Black Studies, v. 18 (1987), pp. 170–190.

10. L. Lorch, *The Painful Path Toward Inclusiveness*, A Century of Mathematical Meetings, AMS, 1996, pp. 83–101.

11. V. Mayes, *Black and Female*, AWM Newsletter, v. 5, no. 6 (1975), pp. 4–6.

12. ___, untitled, AWM Newsletter, v. 18, no. 6 (1988), pp. 8–10.

13. V.K. Newall, J.H. Gipson, L. Waldo Rich, B. Stubblefield (eds), *Black Mathematicians and their Works*, Dorrance & Co., Ardmore, Pa., 1980.

14. *Science*, AAAS, Washington, DC, http://sci.aaas.org/nextwave/print/minorities.

15. *Three Score & Ten: a History of the Southeastern Section of the Mathematical Association of America: 1922–1992.* MAA 1992.

DEPARTMENT OF MATHEMATICS AND STATISTICS, YORK UNIVERSITY, NORTH YORK, ONTARIO, CANADA, M3J 1P3

E-mail address: lorch@mathstat.yorku.ca

DIMACS Series in Discrete Mathematics
and Theoretical Computer Science
Volume **34**, 1997

The Challenge of Diversity*

Etta Z. Falconer

ABSTRACT. The representation of African Americans, Hispanics, and Native Americans in mathematics is well below their representation in the population. The causes of this disparity do not lie within biological or inherent deficiencies in these groups, but in the existence of unfavorable conditions in American society and in the culture and actions of the mathematics community. With dedication and immediate appropriate action, the mathematics community can reconstruct itself into a diverse body with the capability of moving mathematics to greater heights.

The continued vitality of the mathematics community in the twenty-first century depends upon how well diversity can be promoted today. The concept of diversity embraces the acceptance in the workplace, educational institutions and other settings of groups defined in many terms such as race, ethnicity, gender and class. The concept recognizes differences in people who are from different places or backgrounds and promotes acceptance of their differences in appearance, dress, language, religion and other areas.

Already industry is embracing the concept of diversity, because it leads to greater profits. Our country is only a small portion of the global marketplace and we must compete with people from other countries. We no longer have the privilege of thrusting our culture and language upon others. One of the best strategies for gaining entry into a new marketplace is the demonstration of an understanding of the culture and traditions of that place. Companies have found that often this can be better accomplished by bringing into the workforce individuals who were raised in these traditions and culture. This leads to a diverse workforce and in turn, to the achievement of financial goals.

Diversity alone is not enough. We must also aim for equity and inclusiveness. In recent years affirmative action and special programs for women and minorities have been under attack. Affirmative action was established to provide equal access to those groups who had been denied access to the rights of society, such as education and jobs. It was never intended to provide jobs for the unqualified. The current shortage of career opportunities in mathematics is not due to affirmative action policies. Although a small number of qualified women and minorities are

1991 Mathematics Subject Classification: 01A65, 01A80.

*This paper is based on a talk, "The Challenge of Diversity in the Mathematics Community", given at the University of Wisconsin-Madison in May, 1996.

successfully competing for positions, this is not the cause. The mathematics community must analyze the real causes for the shortage of career opportunities for mathematicians and take appropriate action.

Since the late 1960's extraordinary efforts have been made to increase the participation of minorities and women in the scientific and technical workforce. Many programs were initiated, funded by government, industry, community organizations, foundations and professional organizations. The advocates of this movement in the 60's could not have imagined that after 30 years the massive disparity would still exist. And thus lies the challenge. How do we continue to push for a diverse mathematics community at a time when so much remains to be done toward increasing the participation of women and minorities, but efforts in this direction create fear over career opportunities? Lee Lorch [10] in an examination of the progress of the mathematics meetings states, "The position of female and minority mathematicians and the opportunities for members of these communities to become mathematicians are still far short of what they should be. Unemployment affects our successors and we don't know what to do about it."

In the past two years, several court rulings, proposed bills, talks and articles have threaten the efforts to diversify mathematics, science and engineering. The University of California Regents Resolution of 1995 [2] eliminated race, religion, sex, color, ethnicity, or national origin as criteria for admission to the University of California. The California Civil Rights Initiative has been placed on the 1996 Fall ballot. (Proposition 209 passed in November [3].) It prohibits the State from discriminating against, or granting preference to any individual or group in state employment, education, and contracting on the basis of race, sex, color, ethnicity or national origin. As innocent as this seems, its impact was already being felt in 1995. The number of African Americans in the top tier of engineering applicants in California for the 1995-96 class dropped by 40%. But even more significant are attempts to develop similar bills on a national level. The backlash demonstrated here casts doubt on the success of efforts to encourage diversity in mathematics.

Kenneth Ross, in the President's column of *Focus* [17], addressed the apparent reversal of the nation's decades of commitment to civil rights. He stated that the Mathematical Association of America will not alter its commitment to increase minority participation in mathematics at all levels. He further clarified his use of minority to mean underrepresented minorities. The major MAA efforts to increase minority participation in mathematics have been made through the SUMMA Program (Strengthening Underrepresented Minority Mathematics Achievement Program).

It is unfortunate that attempts to reverse the times and minimize gains should come now. Surely, special intervention programs for women and minorities would not have been needed for an extended time, but only until structural changes could be made that would foster equal access. But these structural changes are not yet in place. African Americans, Hispanics and Native Americans are still severely underrepresented in mathematics in the attainment of degrees and in mathematical careers. In spite of women making major gains, they still have not achieved equity. Although parity for men and women with respect to bachelor's and master's degrees in mathematics has almost been attained, as illustrated in Table 1, the percentage of women Ph.D.'s is far too low.

Table 1
Women Mathematics Graduates, 1991

Degree	Women %
Bachelor's	47.8%
Master's	43.1%
Ph.D.	20.1%

Source: Adapted from [14]

Women constitute only 9% of Ph.D. mathematics faculty and hold only 17% of nonacademic mathematics and computer science doctoral positions [5]. Women also make up a significantly larger portion of adjunct and non-tenure track positions in mathematics in higher education. Action must be taken to bring more women into the mathematics community as full partners and a few stars are not enough. Having established the fact that women do not occupy an equitable position in the mathematics community, we will now focus on the severe underrepresentation of certain minority groups and actions that should be taken by the mathematics community to alleviate the situation. In the remainder of the paper, the term "minority" will be used in reference to African Americans, Hispanics, and Native Americans.

Minorities made up almost 22% of the population in 1990, but in 1993 accounted for only 10.6% of the bachelor's degrees in mathematics and only 4.3% of the Ph.D. degrees in mathematics (Table 2). If they had received doctoral degrees in proportion to their population, then 125 doctoral degrees in the mathematical sciences would have been awarded to minorities in 1993 instead of 25.

Table 2
Mathematics Degrees Awarded in 1993

Degree	U.S. Citizens	Minorities	Minority %
Bachelor's	14,074	1,490	10.6%
Master's	2,735	184	6.7%
Ph.D.	583	25	4.3%

Source: Adapted from [13]

1. The changing demographics

Emerging demographic patterns indicate that the health of the mathematics community is dependent upon its success in increasing the participation of minorities. Various reports from major agencies project a significantly different population mix for the next century. Now, more than 20% of the population is minority. By the year 2020, minorities will make up one-third of the population and by 2050, they will constitute 47% of the population. In a few years, Hispanics will replace African Americans as the nation's largest minority. Immigration plays a small role in these enormous demographic changes, but not the major one. The major reason is the higher birth rates of American minorities and the projected continuation of this status for some time due to the younger average age in minority groups.

Changes of the same magnitude are projected for the workforce. According to *Workforce 2000* [9], one in six members of the workforce will be minority by the year 2000. In addition half of all jobs will require some college and most positions will require a significant knowledge of mathematics. Of extreme importance are the new entrants, that is persons of ages 18-24 who will be eligible to enter full-time as workers, college students, or members of the Armed Forces. This is the population that contains most of the college students and which forms the base for our graduate programs in mathematics and mathematics-related areas. But the projections in Table 3 indicate that the population of new entrants will change.

Table 3

New Entrants, 18-24 Years

	1993	2030
African American	13.8%	16.1%
Asian American	3.8%	8.5%
Hispanic	12.3%	21.8%
Native American	0.8%	0.9%
White	69.3%	52.7%

Source: Adapted from [14]

By 2030 there will be a different mix of college students. What do the changes in new entrants mean for the mathematics community? Observant faculty have learned that one must teach the students that enter the institution because there are no more. But the students of the future cannot look the same as the ones of today. It is with some urgency that the community must learn to garner success from the new students and must make every effort to bring them into the community. The future is always with the young, and any nation or community that does not develop its young to replace its members will most surely decline. The mathematics community is no exception.

2. Minorities—still in short supply

As we examine the trends in mathematics, we find that progress for minorities has been painfully slow. According to *Moving Beyond Myths—Revitalizing Undergraduate Mathematics* [4], "College mathematics attracts far too few Black and Hispanic students, and their attrition rates between high school and the sophomore year of college are much too high... Only a small fraction of our population—consisting primarily of white males–complete a mathematics education that matches their potential and interests. The result is an appalling waste of human potential, denying to individuals opportunity for productive careers and to the nation the resources for economic strength."

The number of bachelor's degrees in the mathematical sciences awarded to minorities has shown a moderate increase from 1981 to 1993, but minorities earned 10.9% of these degrees in 1981 and only 10.6% in 1993 (Table 4).

Table 4

Mathematical Sciences Baccalaureates

	1981	1993
U.S. Citizens	10,717	14,074
African American	878	965
Asian American	392	915
Hispanic	275	407
Native American	18	55
White	9,447	11,669

Source: Adapted from [14]

Less we become complacent with the data in Table 4 because of the increase in degrees awarded to African Americans, Hispanics, and Native Americans, recall that minorities constituted 22% of the population in 1993. This means that the equitable share of the degrees is 3,092 instead of 1,490. The estimate of 3,092 is modest given the fact that minority youth constituted almost 27% of the population from the ages 18-24 which is the age group that yields the largest portion of bachelor's degrees.

Although there appears to be no specific data that indicates differences among racial and ethnic groups in attrition from the mathematics major during the college years, it is reasonable to conclude that assumptions that can be made from data taken from all fields (Table 5) are valid in the case of mathematics.

Table 5

Full Time Lower and Upper Division Enrollments at Four-Year Institutions

	Lower Div. 1988 (in thousands)	Upper Div. 1990 (in thousands)	% Decrease
African American	277	163	41.2%
Asian American	104	101	2.8%
Hispanic	169	124	26.6%
Native American	16	12	25.0%
White	2,204	1,874	15.0%

Source: Adapted from [14]

The data in Table 5 indicates that the proportional loss of underrepresented minorities far exceeds that of the White and Asian American population. The same pattern appeared in an examination of similar data for the pair of years 1986 and 1988. Undoubtedly, more attention should be given to the loss of underrepresented students from the lower division to upper division.

Table 6
Master's Degrees Awarded

	1985	1993
U. S. Citizens	2,152	2,735
African American	52	98
Asian American	164	197
Hispanic	53	78
Native American	7	8
White	1,873	2,354
Nonresident	685	1,138

Source: Adapted from [13]

The number of master's degrees in the mathematical sciences awarded to African Americans, Hispanics, and Native Americans increased by 64% from 1985 to 1993, while the growth was 20% for Asian Americans and 26% for White Americans (Table 6). A major improvement has taken place, but the numbers are still too small. To agree with the 22% of the population that minorities constitute, there would have been approximately 600 master's degrees rather than 184.

Significant gains in the earning of doctoral degrees by minorities are not apparent in Table 7. They earned 3.3% of the degrees in the ten-year period 1982-91 and 3.5% in 1992. An average of 13 degrees per year were awarded in the ten year period before 1992, and 16 degrees were awarded in 1992. However, this was during a period of rapid population growth for minorities.

Table 7
Mathematical Sciences Doctoral Degrees

	1982-91	%	1992	%
United States	3,936	100%	452	100%
African American	51	1.3%	4	0.9%
Asian American	143	3.6%	20	4.4%
Hispanic	70	1.8%	10	2.2%
Native American	8	0.2%	2	0.4%
White	3,525	89.6%	405	90%

Source: Adapted from [14]

Table 8 contains data on recent mathematics doctorates which was presented in the AMS-IMS-MAA Surveys of 1993-95. Minorities earned 38 doctoral degrees in the period 1992-95, or less that 3% of the doctoral degrees awarded to American citizens. The highest percent of 2.8% occured in 1994-95. With this data, it is difficult to conclude that any real progress in being made.

Table 8
Recent Mathematics Doctoral Degrees

	1992-93	1993-94	1994-95
U. S. Citizens	526	469	567
African American	7	3	6
Asian American	33	30	24
Hispanic	4	7	9
Native American	0	1	1
White	369	427	527
Others	675	590	640

Source: Adapted from [6], [7], [8]

An examination of the number of minority doctorates in the workforce in Table 9 again points to the severe underparticipation in mathematics that is endemic in the country. Minorities constitute only 4.2% of this group.

Table 9
Mathematical Science Doctorates in the 1991 Work Force

	Number	Percent
Black	200	1.0%
Asian American	2,200	11.1%
Hispanic	500	2.5%
Native American	-	-
White	16,800	85.4%

Source: Adapted from [14]

It is clear that the representation of African Americans, Native American and Hispanics in mathematical doctorates is well below their representation in the U.S. population and also well below their representation in the U.S. labor force. Why do African Americans constitute 10.4% of the workforce and yet only 1.5% of the mathematical doctoral workforce? The data we have examined demonstrates that in spite of the fact that we believe on an intuitive basis that much progress has been made in increasing the participation of minorities in the mathematics community, we cannot justifies this feeling with real data, especially at the doctoral level.

3. The causes of underparticipation

The relatively small number of minorities majoring in mathematics at the undergraduate level cannot be attributed to a lack of interest. One of the indicators of interest in mathematics is the choice of mathematics as a college major. In 1992, African American and Hispanic college freshmen were more likely than majority students to choose a major in mathematics or computer science [14].

Some members of the community believe that the small number of minorities that succeed in mathematics is due to low ability in the thinking skills necessary for success. According to *Moving Beyond Myths*, it is a myth that women, African Americans, and Hispanics are less capable in mathematics [4]. In fact, low ability could only be justified if equal opportunity for learning mathematics had been available. For most minorities this is not the case. "Students enrolled in advanced high school mathematics courses come disproportionately from white upper class and middle class families. Differences in culture and parental expectation magnified by differential opportunities to learn imposed by twelve years of multiple-tracked classes produce vastly different evidence of mathematical power" [11]. At the Project Kaleidoscope National Colloquium, Shirley Malcom [15] stated, "There is no evidence to support biological determinism—that is, any genetic basis for some group's participation or non-participation, performance or achievement in science and mathematics. Complex social and cultural factors, opportunity to learn, and the availability of resources have likely interacted to produce the current distribution of groups in science fields."

With the dismissal of lack of interest and low ability as causes of the disproportionate low number of minorities in mathematics we must continue to look for the true reasons. Certainly, inadequate academic preparation plays a role because minorities are less likely to attend the schools that offer the best preparation for college mathematics. A smaller portion of college-bound African American seniors that White Americans complete geometry, trigonometry and calculus. White students have the same advantage with respect to Native American students. Although inadequate academic preparation is one of the reasons that minorities leave mathematics, particularly at the undergraduate level, it is not the only one. Educational environment is a more significant factor. This provides hope for we can do more to control the educational environment than major societal systems which provide advantages to certain socioeconomic classes. We should note that lasting change, however, requires that we also find solutions to the latter.

4. The effect of culture shock

Most minority children attend public high schools in urban areas where they are in the majority. For example, minorities constitute more that 80% of the public school population in Washington, Newark, Chicago and San Antonio. Most minorities attend predominantly white colleges. Therefore, the transition from high school to college is also a transition from a setting in which the student is in a majority position to one in which the student is in a minority position. Moving into this new environment with its different culture is stressful for minority students who must also adjust to the common stresses that all students experience in changing from high school to college.

Minorities must spend time and energy learning different patterns, values and behaviors in college while majority students can devote this same amount of time and energy to their studies. My own experiences at the University of Wisconsin serve to illustrate this. I entered UW-Madison in 1953 to pursue a master's degree in mathematics with the encouragement of Dr. Lee Lorch, Chairman of the Mathematics Department at Fisk University in Nashville. Can you imagine what it was like for a 19 year old Black female from Tupelo, Mississippi who had been immersed in segregation for all of her life to go to the University of Wisconsin? I underwent a major culture shock. I was culturally and racially different from every classmate and every professor. Thrust in the midst of total strangers and missing a feeling of belonging and comfort, I sought contact with others on a nonacademic basis in order to survive. Some were students in other fields and some were people in the local community. I gravitated to students from Africa, a roommate from Thailand, and an office-mate from India, who was the only person to whom I could ask a mathematics question. I followed a friend in political science to political discussions with Jewish students. The situations that I had previously experienced with white Americans were very limited and except for the white faculty at Fisk, mostly negative ones. My relationships at the University of Wisconsin firmly shaped the closeness that I feel toward people who are different from me, especially those from other countries.

Dr. R. Creighton Buck was my academic advisor and helped me to select my courses. Dr. C. C. MacDuffee was chairman of the department. He was so kind and helpful to me. We used his book in his class and I worked very hard for him. I was very fortunate. I did not realize at that time that there were so few Negro women studying mathematics. I only knew that there was only one at Wisconsin. If you can understand my culture shock, then you must also understand that the department was well ahead of the times. It had welcomed a young Black woman into the graduate program and had entrusted her with teaching white students through a teaching assistantship. In spite of the fact that I did not have graduate students with whom I could study and discuss mathematics, I was not alienated. The University of Wisconsin not only gave me a strong education, but exposed me to many things and different people, thus playing an important part in my growth as a person.

However, many minorities students do feel alienated on campus. Alienation results when there is a very small number of minority students overshadowed by a large number of majority ones. Alienation has been directly related to attrition of African American students from college. It is not a minor factor, but its effects can be minimized with appropriate action. There is a tendency to treat minorities as "invisible". Everyone is quite aware that the minority student is present in the class but efforts are made not to recognize it. Not long ago, I served on an accrediting team for a high school in the metropolitan Atlanta area and visited the AP calculus class. This class was obviously being taught at great sacrifice for the school since it contained only eight students, seven majority and one African American male. This student sat on the periphery of the group next to a white male. The white male teacher taught in an interactive style, with questions to the students. He knew their names and called on them by name. Not once during the period did he call on the Black student—not even when the student raised his hand in response to a general question posed to the class. The teacher did not look in the direction of the student during the entire period. The student was in the class, but effectively invisible.

One would think that this would not happen at the graduate level, but about a year ago a situation was described to me that indicates that it does. There was one African American male in a small graduate mathematics class with majority and foreign students. It was the pattern of the white male professor to never look at the Black student during the class. The student was effectively invisible. The student endured this constant insult and tried to achieve. One day, only one white student and the Black student came to class. The professor pulled up a chair directly in front of the white student and proceeded to carry on a discussion of the material while the African American student sat in the room alone, trying to absorb whatever he could from the discussion from which he was excluded. For minorities it sometimes requires extraordinary ability and perseverance, because the system sets up incredibly powerful barriers.

From my own teaching experience, I have learned about the importance of self-confidence and self-esteem in achieving student success in mathematics. As a result I spend some time in class and in student conferences on building self-confidence, because the use of this time pays at least as many dividends as discussing mathematics content. It is difficult to build self-confidence in the presence of a culturally held stereotype that minorities cannot do mathematics, or anything else that requires higher thinking skills. All faculty should be aware of this, and work to minimize its destructive effects.

Role models and mentors are important contributors to the success of minority students. At Fisk University I was fortunate to have both. My role model was my teacher, Dr. Evelyn Boyd Granville, one of the first two African American women to earn the Ph.D. in mathematics (Yale, 1949). Dr. Lee Lorch, my undergraduate chair and teacher is my mentor even until now. White males can be excellent mentors. At my institution, Dr. Jeffrey Ehme, a young white professor in the Mathematics Department, can be credited with mentoring an African American female student who earned the first prize in the poster section at the Joint Mathematics Meetings in January, 1995. She is now pursuing a graduate degree in mathematics. Until he provided her with close attention and encouragement she was floundering in her college studies. Majority mathematicians at predominantly majority institutions can have the same influence on minorities. It is unnecessary and impractical to draw mentors only from the minority professoriate.

Campus climate—the totality of the aspects of institutional atmosphere and environment which foster or impede minorities' academic, social, personal and professional development—has been identified as a major factor in the success of minority students. Issues related to campus climate include classroom and out-of-classroom experiences, financial aid, student services, and advising. According to a report by Dr. Howard Adams [1], a chilly climate that is unsupportive and fails to nurture students is the greatest inhibitor to the participation and success of engineering and science students, especially minorities and women.

My former students often share their graduate school experiences with me. Two recent Spelman College graduates were allowed to register in courses for which they did not have the proper background. They followed published materials that listed requirements for the degree which caused difficult and discouraging situations. One left her graduate program after completing the master's degree. The other student is now at dissertation level, but still bears the psychological scars.

Another climate-based problem that minority students experience is the informal way in which some information is communicated in graduate school. Isolated minority students may not be in the communication network of other graduate students. Without access to the usual communication lines, minority students may behave in ways that are interpreted by faculty as expressions of disinterest in mathematics, or worse—hostility.

Another challenge is low expectations of minority students by faculty. A student usually knows whether a faculty member expects him or her to achieve. The lack of high expectations causes limited performance and even failure. Standards should never be lowered for any student. Work should be demanding and efforts should be made to keep up student morale. One of my best students entered a good university last year. A faculty advisor told her and another minority student that they should enroll only in undergraduate courses because they really did not quality for admission and had been accepted on the basis of affirmative action. She was devastated. If the student needed to take undergraduate courses, then this could have been accomplished without implying that she was unworthy. As a result, she left the university before the semester ended. We have succeeded in getting her into a more supportive university and believe that she will be successful.

Minorities do quite well in mathematics at some institutions. The institutions in Puerto Rico have been exporting scientists for some time. The Historically Black Colleges and Universities (HBCU's) have been extraordinarily successful in producing mathematics baccalaureates in spite of a smaller financial base, more needy students and more students with inadequate academic preparation. What could account for this? The primary factor is the nurturing and accepting campus climate. HBCU's are quite challenging. Expectations are high and students are held accountable for learning. In contrast, some of the brightest minority students are recruited to predominantly majority institutions, only to become victims of the system and forever lost from mathematics.

A study of 1,100 institutions of higher education, made by Carol Fuller for Project Kaleidoscope, indicated that HBCU's are a significant baccalaureate source for science Ph.D.'s earned by African Americans [16]. Predominantly minority institutions and other institutions in the southwestern United States, Florida, California and Puerto Rico were the baccalaureate source of one-half of the Hispanic doctorates in science. She found that in 1989, only 279 institutions awarded a bachelor's degree in mathematics to an African American, only 158 awarded one to a Hispanic, and only 36 awarded the degree to a Native American. In other words, 75% of the institutions did not award a single mathematics bachelor's degree to an African American, 85% did not award a mathematics bachelor's degree to a Hispanic student, and 97% did not award one to a Native American. With the failure of so many mathematics departments to participate even minimally in the production of minority mathematics baccalaureates, it is a wonder that 1,210 degrees were awarded that year.

There are valuable programs such as MAA's SUMMA which will increase the number of doctorates in the mathematical sciences and the number of minority mathematics faculty. However, these programs are not the total solution to the creation of a diverse mathematics community. This does not mean that the programs are not essential. We will need special programs until mainstream programs work to serve all students better. Shirley Malcom [12] in a discussion of policies to promote change noted that the development of strategies to draw the talents of all Americans into science and engineering will require commitment by the larger communities of professionals in these fields, by the institutions that educate and hire, and by the larger science policy structures which direct the flow of resources.

We can expect improvement to occur in K-12 mathematics programs. Already, there are large state, urban and rural systemic programs underway. In addition, the new national standards in mathematics are being adopted in some form by many states. Structural change at all levels, coupled with special programs will eventually lead to the goals of excellence and diversity. There are some things that mathematics departments can do now that would help to bring about this change.

Mathematics departments should:

1. Establish a level of commitment to having minority students as undergraduate majors and graduate students, as well as having minority faculty. This commitment should be widely communicated to faculty and students.

2. Establish goals for minority undergraduate and graduate students. For a majority of mathematics departments, a determination to produce even one minority bachelor's degree a year would be a major improvement.

3. Mobilize all faculty, staff and students to provide a supportive environment for minority students and faculty.

4. Develop appropriate language of encouragement that is used publicly and by all persons. The language should make it clear that minority students are in the department for the same reason as other students—because they have the potential to do well in mathematics.

5. Extend first-class citizenship to all minority students and faculty. In particular, give equal access to the available teaching and research assistantships in the department. Equal access does not mean welcoming minority students to apply for financial assistance and always selecting them for grants or scholarships which provide financial support but do not integrate students into the department and do not provide the student with the networking and other experiences that lead to productive research after graduate school.

6. Make certain that minority graduate students are included in the informal information system. The department should inform the faculty of its responsibility to communicate at the same level with all students. Majority students should not be the only ones receiving information that is essential for moving through the system. Graduate minority students need a proper dissertation advisor, a proper committee, the right courses, discussions of mathematics over coffee, the goodwill fostered by attending the department mathematics talks, a knowledge of actions which promote and those which retard progress, healthy research connections with faculty, help in submitting a paper to a journal, and suggestions of mathematicians to send advance copies of a research paper. In other words, they need the same things that have proven effective in moving other students into mathematics research.

7. Provide a climate in which minority students are encouraged to achieve academically. Undergraduate research is an excellent way to encourage any student to move forward in mathematics.

8. Encourage faculty to acknowledge minority presence; making students invisible should not be tolerated. Minority students should be looked at and called upon within a professor's style, just like other students. All students need feedback and encouragement.

9. Devote department meetings or workshops to discussing what the department can do to improve its outcomes with minority students and faculty. Visible successful outcomes should be required. Discussions should include the climate in the department and how it can be improved, informal networking and how the minority student can become better connected to the department, mentoring and who will become mentors, and attitudes and how they can be changed. Yes, African Americans can do good mathematics research in an open mathematics department.

10. Initiate a program that improves advising. For undergraduate and graduate students, academic advising is an individual affair—not just a matter of having a student follow a written guide. Entering minority graduate students may not be able to assess whether they need a bridge course or the first graduate course in analysis. Discussing what students have covered may prevent the unfortunate situation of having students in courses without proper preparation, or the worst situation of having them sit through material already learned.

These suggestions are not exhaustive and they are not intended only for mathematics departments at predominantly majority institutions. There are many predominantly minority institutions, including HBCU's, that need to address the dismal production of minority graduates from their mathematics departments.

Although achieving diversity is a major problem for the mathematics community, I have great faith. The mathematics community contains some of the best minds in the world and its people have the best problem-solving abilities in the world. There are many members of the community, both minority and majority, who are just as concerned about the plight of minorities in mathematics as I am. We need everyone to join in the effort to broaden the base. Then, we will no longer be a community of a privileged class. The mathematics community in America will be a mosaic representative of the diversity of its population. At that point, mathematics itself will begin to soar to heights never before conceived.

5. References

1. H. Adams, *Creating a Campus Climate That Supports Academic Excellence*, The GEM Program, Notre Dame, 1994.

2. M. Barinaga, *Showdown at the UC Corrall*, Science, 271 (5250), Feb. 9, 1996, p. 752.

3. —, *California Bans Affirmative Action*, Science, 274 (5290), Nov. 15, 1996, p. 1073.

4. Committee on the Mathematical Sciences in the Year 2000, Board on Mathematical Sciences, Mathematical Sciences Education Board, National Research Council, *Moving Beyond Myths: Revitalizing Undergraduate Mathematics,* National Academy Press, 1991.

5. Cross University Research in Engineering and Science, *The Equity Agenda: Women in Science, Mathematics and Engineering*, The Center for the Education of Women, University of Michigan, 1996.

6. J. Fulton, *Report on the 1993 Survey of New Doctoral Recipients,* Notices American Math. Soc. 40 (1993).

7. —, *Report on the 1994 Survey of New Doctoral Recipients,* Notices American Math. Soc. 41 (1994).

8. —, *Report on the 1995 Survey of New Doctoral Recipients,* Notices American Math. Soc. 42 (1995).

9. W. Johnston and A. Packer (Eds.) *Workforce 2000 - Work and Workers for the 21st Century*, The Hudson Institute, 1987.

10. L. Lorch, *The Painful Path Toward Inclusiveness*, A Century of Mathematical Meetings, AMS, 1996, pp. 83-101.

11. Mathematical Sciences Education Board, Board on Mathematical Sciences, Committee on the Mathematical Sciences in the Year 2000, National Research Council, *Everybody Counts: A Report to the Nation on the Future of Mathematics Education,* National Academy Press, 1991.

12. M. Matyas and S. Malcom (Eds.), *Investing in Human Potential: Science and Engineering at the Crossroads*, AAAS, 1991, pp. 139-145.

13. National Science Foundation, *Science and Engineering Degrees, by Race/ Ethnicity of Recipients* 1985-93, NSF, 1995.

14. —, *Women, Minorities, and Persons with Disabilities in Science and Engineering: 1994*, NSF, 1994.

15. Project Kaleidoscope, *What Works: Building Natural Science Communities. A Plan for Strengthening Undergraduate Science and Mathematics*, Vol. I, PKAL, 1991, pp. 25-27.

16. —, *What Works: Building Natural Science Communities. A Plan for Strengthening Undergraduate Science and Mathematics*, Vol. II, PKAL, 1991, pp. C99-C129.

17. K. Ross, *President's Column*, Focus, MAA, Oct. 1995.

DEPARTMENT OF MATHEMATICS, SPELMAN COLLEGE, ATLANTA, GA 30314
E-mail address: efalcone@spelman.edu

DIMACS Series in Discrete Mathematics
and Theoretical Computer Science
Volume **34**, 1997

What Next?
A Meta-History of Black Mathematicians

Patricia Clark Kenschaft

ABSTRACT: This article reviews the sources of published information about black mathematicians prior to the past decade and suggests avenues of research.

1. Introduction and Overview

As a field of study, the History of Black Mathematicians is off to a good start, but it lacks coherence. These symposia "prove" that the topic is of interest to far more than those struggling against enormous odds to have a rewarding career; financial support and AMS cooperation show that others are concerned. The past decade has yielded an abundance of short piecess about black mathematicians in the AMS *Notices*, the MAA *Focus*, the encyclopedia *Black Women in America*, various NAM publications, the *Chronicle of Higher Education, Black Issues in Education,* the newsletters of the Benjamin Banneker Association and AWM, and the recently deceased but once pervasive *Undergraduate Mathematics Education (UME) Trends.* It would be useful to collect these articles and pieces and publish them in one volume, although an editor might want to request newly written material to fill obvious gaps.

However, much more remains to be done, and some groundwork is in danger of being forgotten. This article suggests directions for further research and describes five books that have been out of print for some time but may be useful to researchers. The surfeit of publications about why blacks "don't" do mathematics will be ignored. In 1986 and 1987 I did a survey of black mathematics of New Jersey with virtually no support and only one course of research time (a quarter load, while teaching three other courses) for two semesters. By networking, beginning with about ten people I already knew, I eventually identified 175 black mathematicians in my own state, where "mathematicians" are defined to be people who either work in or have adegree (at any level) in mathematics. My work was incomplete, and I suspect there were at least double that number.

If New Jersey is typical, then there are over twelve thousand black mathematicians in the country. That is not a large proportion of the black population, but it is a significant number of people to publicize, serve, and mobilize. The repetitive, depressing question about why blacks "don't" excel at mathematics could be effectively refuted if our society acknowledged the many who do. Finding them will be a major project for avid researchers with interviewing skills.

1991 *Mathematics Subject Classification.* 01A02

At the doctoral level, there are already lists, currently kept at the MAA SUMMA (Strengthening Undergraduate Minority Mathematical Achievement) office. Surveys of people on these lists could yield statistical trends among the super-achievers. Several of them merit complete biographies along the lines of Kenneth Manning's *Black Apollo in Science: the Life of Ernest Everett Just* (New York City: Oxford University Press, 1983). A biography of Elbert Cox is in process.

A historian who enjoys archives could investigate the free African school, Institute for Colored Youth (the predecessor of Cheyney State University), near Philadelphia in the pre-bellum nineteenth century about which, apparently, nothing has been written recently. The library and archives of any of the traditionally black institutions would yield information about black mathematicians, and the Howard University mathematics department could be the subject of a fascinating book. The National Association of Mathematicians, and its officers, is another worthy topic.

An incomplete set of recently published articles can be obtained by requesting them from me. My own most ambitious publications in this vein are "Black Women in Mathematics in the United States," in the *American Mathematical Monthly* (October, 1981, 88:8, 592-604, "Black Men and Women in Mathematical Research" in the *Journal of Black Studies* (December, 1987, 19:2, 170-190), and eight entries in the *Encyclopedia of Black Women: An Historical Perspective* (Carlson Publishing, Inc., 1993). These are some of the longer prose pieces extant about black mathematicians, but they hardly begin to say what needs to be said.

Claudia Zaslavsky has written books about African mathematics based on her travels in Africa, most notably *Africa Counts: Number and Pattern in African Culture* (Chicago, IL, Independent Publishers Group, 800-888-4741, 1973) and *The Multicultural Math Classroom: Bringing in the World* (Portsmouth, NH, Heineman, 800-541-2896). A more recent book is *Math: A Rich Heritage* written collaboratively (Melville, NY, Globe Fearon, 800-872-8893, $4.95), which highlights the contributions of African Americans whose backgrounds helped them succeed in science, technology, and other mathematics-related careers.

2. Out-of-Print Books

Holders of Doctorates Among American Negroes (1876-1943) by Harry Washington Greene (Boston: Meador Publishing Co, 1946) includes information about 368 of the 381 blacks the author knew to have earned a doctorate of "equivalent" before December 1943. Eight were in mathematics. Green omitted Joseph Alphonso and Clarence Stephens.

The Negro in Science by Julius H. Taylor (Baltimore: Morgan State College Press, 1955) reprints some papers, along with names of leading scientists. There is considerable commentary, the tone of which is indicated by the passage, "It is certainly arguable, however, that the very intensity of Just's immersion in research was an attempt to escape, through his reputation as a scientist, the status imposed by being born a Negro. This is so obvious a means of compensation that one wonders that there have not been more Negroes literally working their hearts out to establish their independence though unmistaken mastery of some respected segment of learning."

Mathematicians included in this book other than those on the list below are Andrew Aheart, Ethelbert William Haskins, Nan Phelps Manuey, Georgia Caldwell Smith, and George Frederick Woodson, Jr.

The Negro Woman's College Education is not primarily about mathematics, but it does shed light on our subject. It is a highly readable summary of Jean L. Noble's doctoral dissertation (NYC: Bureau of Publications, Teachers College, Columbia University, 1956). The author sent a survey to 1000 college-educated black women in six metropolitan areas and analyzed the 412 replies.

Noble quotes Mary Church Terrell (A.B. Oberlin, 1884) saying in 1953, "I was ridiculed and told that no man would want to marry a woman who studied higher mathematics. I said I'd take the chance and run the risk." Terrell's expressed regret in her autobiography for entering a fine marriage because it required her to turn down an offer to become registrar at Oberlin College, which would have set a precedent for hiring black faculty at predominantly white colleges. "If I acted unwisely, I am sorry... I certainly deprived myself of... being the first and only colored woman in the United States to whom such a position has ever been offered, so far as I am able to ascertain," she wrote in 1940. [*A Colored Woman in a White World*, Mary Church Terrell, Washington, DC, Ransdell, Inc., 1940, p. 103]

Noble tells us that in 1920 only 20% of the graduates of "Negro colleges" were women, but by 1953 only 35.6% of the degrees went to men. She hypothesizes that black women tried especially hard to attend college because secretarial work and store clerking, the major paths for white women out of menial work, were closed to black women. In 1950 women comprised 58% of all black professional workers. She reported that 1.6% of all black men were college graduates, 6% of the white men, 2.1% of black women, and 4.4% of white women.

In *Negroes in Science: Natural Science Doctorates, 1976-1969* (Detroit, Box 790: Belamp Publishing, 1971), James M. Jay reports statistical studies of 587 blacks holding doctorates in biological, medical, physical, pharmaceutical, or agricultural science, but does not give any names. Of his group, 58 (10%) were women, about 5% were deceased, and about 25 to 30 people per year were joining the group.

He observed that the average number of years elapsing between undergraduate and doctoral degrees had not changed over the past decade and was about three years more than the average of 7.6 for all natural science doctorates between 1920 and 1961. He wrote, "While large numbers of Negroes left the south since 1930, the proportionate increase in science doctoral recipients of northern origin has not taken place. The increases are taking place in the south even at a higher rate." (p. 9) A doctoral dissertation is in the planning stages about the origins of scholarly mathematicians and the impact of segregated vs. integrated education.

Black Mathematicians and Their Works by Virginia Newell, Joella Gipson, L. Waldo Rich, and Beauregard Stubblefield (Ardmore, PA: Dorrance and Company, 1980) includes information about 62 blacks in mathematics with great variation in the completeness of the biographies. There are numerous photographs. It reprints one research paper each for 24 mathematicians.

An appendix includes published letters from the 1950's documenting and protesting exclusion of blacks from mathematical meetings. Its "Forward" by Wade Ellis, Sr., includes statements such as, "Nowadays our promising youth are... menacingly threatened by exposure to teachers who... have been convinced to their very viscera that Blacks... are abysmally and irrevocably hopeless as far as mathematics is concerned."

3. U.S Black Mathematicians Who Earned Doctorates by 1955

Twenty-two black people in the United States, including two women, appear on the list below of those who received a doctorate in mathematics before 1955. Twenty-one are mentioned in at least one of the above books.

Albert Turner Bharucha-Reid (1927-1985) did not actually put his research together and submit a thesis, because he didn't see any need to waste time that way in 1953. He was concluding three years of graduate study at the University of Chicago, during which he produced eight published research papers. Until the end, his career did not suffer from the lack of a doctoral degree; he was the dissertation supervisor of thirteen doctoral candidates at Wayne State University, one of whom was black. He published over seventy research papers and was elected to the Council of the American Mathematical Society just before his untimely death. His life could surely be the subject of a provocative biography; it seems strange to omit him from the list below. (Two recent presidents of the American Mathematical Society, Garrett Birkoff and Andrew Gleason, also never actually completed the requirements for a docotral degree for similar reasons.)

For each entry below, I include (in parenthesis) the references where that person is mentioned (Greene, Taylor, *Black Mathematicians and Their Works*, and my own 1987 paper), the birth date if known, the institution that awarded the doctorate, the year the doctorate was received, and the death date if known. The rich heritage already existing for blacks in mathematics is evident after even a casual reading of the list. Many more blacks have earned doctorates in mathematics in the past 40 years.

Cox, Elbert Frank; (Greene, BMW, Kenschaft) 12/5/95, Cornell University, '25, d. 11/28/69
Woodard, Dudley Weldon; (Greene, BMW, Kenschaft) 10/3/81, University of Pennsylvania, '28, d. 7/1/65
Claytor, W. W. Scheiffelin; (Green, BMW, Kenschaft), University of Pennsylvania '33, d. 1967
Talbot, Walter Richard; (Greene, BMW, Kenschaft), 1909, University of Pittsburgh '34, d. 1977
McDaniel, Ruben; (Green, Taylor, BMW), 7/27/02, Cornell University '38
Pierce, Joseph Alphonso; (Green, Taylor, BMW) 8/10/22, University of Michigan, '38, d.
Blackwell, David; (Green, Taylor, BMW, Kenschaft) 4/24/19, University of Illinois, '41 (Fellow of the National Academy of Sciences)
Wilkins, J. Ernest, Jr.; (Greene, Taylor, BMW, Kenschaft), 11/27/23, University of Chicago, '42 (Fellow of the National Academy of Engineering)
Stephens, Clarence Francis; (Taylor, BMW), 8/24/17, University of Michigan, '43
Brothers, Warren Hill; (BMW) 1/15/15, University of Michigan, '44
Dennis, Joseph J.; (BMW) 4/11/05, Northwestern University, '44, d. 1977
Ellis, Wade, Sr.; (Taylor, BMW) 1/9/09, University of Michigan, '44
Certain, Jeremiah; (Taylor) 6/6/20, Harvard University, '45
Granville, Evelyn Boyd; (Taylor, BMW, Kenschaft) 5/1/24, Yale University, '49
Browne, Marjorie Lee; (BMW, Kenschaft) 9/9/14, University of Michigan, '49, d. 10/19/79
Butcher, George Hench; (BMW), University of Pennsylvania, '50
Love, Theodore Arceola; (BMW) New York University
Bell, Charles; 1928, (Kenschaft) Notre Dame University, '53
Mishoe, Luna Isaac, (Taylor, BMW) 1/5/17, New York University, '53
Shabazz, Abdulalim A.; (BMW) 5/22/27, Cornell University, '55
Williams, Lloyd Kenneth; (BMW) 10/6/25, University of California at Berkeley, '55

DEPARTMENT OF MATHEMATICS AND COMPUTER SCIENCE, MONTCLAIR STATE UNIVERSITY, UPPER MONTCLAIR, NJ 07043
E-mail address: kenschaft@math.montclair.edu

DIMACS Series in Discrete Mathematics
and Theoretical Computer Science
Volume **34**, 1997

A Personal History of the Origins of the National Association of Mathematicians' "Presentations by Recipients of Recent Ph.D.'s"

Donald M. Hill

Abstract. This article details the background of the establishment of the Browne-Granville Presentations, an annual occurrence sponsored by the National Association of Mathematicians at the Joint Winter Meetings. They were originally known as the "Presentations by Recipients of Recent Ph.D.'s."

1. Introduction

The National Association of Mathematicians (NAM) is a group devoted to furthering mathematics among the African American community. Since 1989 its "Presentations by Recipients of Recent Ph.D.'s" has been an annual feature on NAM's program at the Joint Winter Meetings. By looking at the origins and initial results we may see the state of African American participation at the doctoral level during the period 1983 - 1987 and some of the efforts developed to strengthen bonds among the new doctorates, encourage them to stay active in research, and participate in professional organizations. Other efforts to increase and maintain minority interest in mathematics are mentioned in passing. Finally, portions of correspondence from department chairs to me will further highlight the dire situation facing the American mathematical community in the production of African American mathematicians. All the information is what was known to me in 1988.

2. The Search

In 1972 I was a new faculty member at Florida A&M University (FAMU), the Historically Black University in the State University System of Florida. I looked forward to comparing new experiences with those I had had teaching mathematics in French in the Congo from 1965 to 1967. As an undergraduate and graduate student in both northern and southern universities I could count on my thumbs the number of my classmates and teachers who were Blacks. Certainly I would find several Black Ph.D.'s in mathematics at FAMU. Wrong. Among our sixteen faculty members there was exactly one. How could this be? My Congolese students had been extremely gifted. Our school was so isolated that it was four days by truck to the nearest doctor and the students had to walk up to 200 miles to attend our boarding school. I often had the only book in class and our Congolese director rationed the chalk at two pieces per day per teacher. Yet even under those conditions I had taught calculus to them as juniors in high school and I would willingly have

1991 *Mathematics Subject Classification.* Primary 01A65.

matched them against the Ivy League students I had taught as a TA before going to Africa. There was no way it could be a question of ability. So, where were the Black Ph.D.'s in American mathematics?

It didn't take long to discover that there weren't many. Then I learned of a truly wonderful group called NAM that had been formally incorporated in 1972. In NAM's statement of purpose an invitation was extended to "any and all persons expressing a serious interest in helping to implement solutions ...," and so I joined NAM.

Believing we should start with what we had, I looked for ways to increase interest in mathematics. I began where I was by including math problems in my regular classes at FAMU that I had used in Africa. I talked with my students briefly about Africa, the outstanding students I had taught there, and their great desire for knowledge. I had to wait many years for someone to invent a name for what I had been doing. Some now call it Afro-centric education.

Another idea that occurred to me was that I should take our FAMU Math Club to the Winter Meetings. We had bake sales, car washes, etc., and when that wasn't enough, various members of the national mathematical community provided donations to enable us to make the trips. I held seminars before the meetings to acquaint the students with the professional organizations and their leaders. We also studied the program carefully to choose which sessions we would attend. A highlight was always meeting the leaders of NAM and the Mathematical Association of America. In 1986 and 1988 pictures of the students with MAA officers appeared in "FOCUS."

Both these activities were well and good, but by now I had been at FAMU for fifteen years and was clearly seeing a disturbing pattern as new Black Ph.D.'s joined other departments at FAMU. FAMU was naturally proud of these individuals and found leadership positions for them as administrators, directors, etc. In other words, our best and brightest were being removed from their disciplines and what little energy and time remained were devoted to teaching with very little left for research and professional involvement. Furthermore the reward system of tenure, promotion, and salary increases tended to move them in that direction. Conversations with new Black Ph.D.'s at majority institutions led me to believe that they too were being pulled away from research and professional activities by being placed on so many committees within their universities.

Somehow we needed to get the new Black Ph.D.'s together at professional meetings so that they could form strong bonds among themselves and with leaders of NAM and other professional organizations. By now I had been elected to NAM's Board of Directors and was editing the "Proceedings" of NAM. I approached the Board with the following:

> New Black Ph.D.'s in mathematics will be identified as they receive their degrees. For the January joint meetings those whose doctoral age will be 12 - 24 months will be invited through the auspices of NAM to give short talks about their dissertations or any other research area of interest to them. They will prepare 4 -8 page summaries of their talks which will be published as part of the NAM "Proceedings." The doctoral age can be more than 24 months for the first time.

I added several comments to the main proposal. Encouragement was to be provided by NAM, MAA, and AMS members for these new Ph.D.'s to become/stay involved with professional activities. They were to meet successful role models and also form an age cohort of their own. It was felt that an invitation to speak at a national meeting would be attractive to them at that stage of their careers as promotion and tenure loomed ahead. The talks were to be from 8 - 12 minutes. The papers were to be placed with accounts of other NAM presentations, photocopied, spiral bound, and sent to NAM members. Extra copies were to be made available to others at a nominal cost. National MAA and AMS officers could attend the talks and then perhaps consider the speakers for committee assignments or other opportunities with their organizations. Sectional officers of MAA could do the same and also consider the speakers as presenters at their sectional meetings.

The NAM Board of Directors enthusiastically supported the idea and I stipulated that all expenses were to be paid by me. With no secretarial help available at FAMU, I decided to buy my own computer system thereby becoming the first person in my department to have a PC. At Sears I bought an Amstrad computer, monitor, keyboard, printer and word processor application for $288. One of our children was in college and the other one headed that way. Money was scarce. While waiting for my next paycheck I taught myself how to use the Amstrad. After the next payday I was able to purchase a list from AMS of all units (departments) offering doctoral degrees in the mathematical sciences in the US. I also bought several boxes of envelopes and two rolls of stamps.

The identification of all Blacks receiving doctorates in the mathematical sciences between 1983 and 1987 was much more difficult than I had rather naively assumed it would be. There were roughly 185 units awarding doctorates and my first letter to them simply asked for the names, and if possible, the addresses of all such Blacks. I enclosed information about NAM and said we wanted to invite them to give a talk at our national meeting. I thought I would get at least a 95% return rate. After all, one didn't need to be a member of the National Academy of Sciences to figure out who the desired persons were. After waiting five weeks I had a return rate of only 60% and I was decidedly unhappy. It was only months later that I learned that was an outstanding level of participation. My second letter to the non respondents was simply a reminder and enclosed a copy of the first correspondence. Again, for my records I asked to hear from the departments whether they had such individuals or not. About 60% of these responded by six weeks after the second letter. That left me with about 40% of 40%, i.e. 16%, of the original 185 units who, for whatever reason, had sent me nothing. By now I was really disturbed with those 30 hard-core units. Thinking that additional letters would produce nothing and running short of time, energy, and money, I chose a more vigorous course of action. I wrote directly to the presidents of those institutions. I explained my chagrin that two attempts had resulted in nothing and that some might view this as indifference, incompetence, or even racism. I asked that the presidents personally write me on their own letterhead. I concluded by writing that it was hard to imagine what combination of unlikely events could have led to my receiving nothing, but that strange things do happen. If such occurred, I asked them to accept my apology to them and their departments.

That got big time results and I now had responses from all but two units, an incredible overall return rate of 99%. I also received more information than I wanted. A couple of letters really tore into me. One, in particular, took me to task for not realizing their program had not been in existence long enough to produce any doctorates. Since I had asked them to let me know even if they didn't have any Black doctorates, it didn't seem to me to be really that difficult a problem. If you have no doctorates, then how many Black doctorates are there?

After months of work, a fair amount of expense, and nine revisions of the list I finally had 41 names. Of the 41 doctorates, 34 resided in the US and 19 of them returned biographical information I had requested in my letter to them inviting them to speak at the NAM meeting in Phoenix of 1989. Of the other 15, five names were received so late that I could not invite them, one had my letter returned "address unknown," and one I simply missed. The others simply could not be reached or chose not to reply.

3. The Presentations

Eight accepted NAM's invitation to speak at the inaugural "Presentations by Recipients of Recent Ph.D.'s." The seven who actually spoke were De Juran Richardson of Lake Forest College (A Parametric Group Sequential Procedure for Comparing Survival Distributions of Two treatments), Arouna Davies of Prairie View A&M University (A Polynomial Time Algorithm and Combinatorial Problems), Nathaniel Dean of Bellcore (Matching Extendibility and the Genus of Graphs), n'Ekwunife Muoneke of Prairie View A&M University (Invariant Perron-Frobenius Normal Form of Nonnegative Irreducible Matrices by a Special Class of Simultaneous Permutations of Rows and Columns), Melvin Currie of the University of Richmond (Topological Implications of Metric Properties), Shiferaw Berhanu of Temple University (Microlocal Holmgren's Theorem for Certain Hypo-analytic Structures), and Abdulkeni Zekeria of Fitchburg State College (Singularly Perturbed BVP with Discontinuous Coefficient). I also wrote letters to all the chairs of the 185 doctoral units inviting them to the session and thanking them for collectively helping me locate the individuals.

The presentations were well received. Obviously many NAM members were there and there was a good representation of leadership persons from AMS and MAA. I collected the papers of the speakers, added to them the papers from other NAM sessions, photocopied and spiral bound all of them, and then mailed these "NAM Proceedings." By devoting my entire spring break to the task, NAM members were able to have the "Proceedings" less than ten weeks after the Phoenix meetings. The NAM Board of Directors decided to make the event an annual occasion and has since 1989 always had "Presentations by Recent Ph.D.'s" as an integral part of the winter meeting. It was recently renamed the Browne-Granville Presentations in honor of Marjorie Lee Browne and Evelyn Boyd Granville, the first African American women to receive doctorates in research mathematics.

4. What Was Known to Me about Them in 1988

Here are some interesting facts. Seven of the Ph.D.'s returned to their home countries of Ethiopia, Ghana, Jamaica, South Africa, Swaziland, Zaire, and Zimbabwe. Of the 19 Ph.D.'s residing in the US who provided biographical information, seven received their undergraduate degrees abroad, six went to Historically Black Institutions, and six attended majority institutions. Five of the 19 were employed in non-academic work (AT&T Bell Labs, Bellcore, BDM Corporation, General Dynamics, and McDonnell Douglas). Five taught at Historically Black Institutions and the other nine were at majority institutions.

Howard University produced seven of the 41 Ph.D.'s, followed by UC Berkeley with five, while Northwestern, Old Dominion, and Southern Illinois each had two. All 41 known to me at that time are now listed with the doctoral granting institution and, when known to me, the year. Darry Andrews (UC Berkeley, 85), Stella Ashford (LSU), Richard Baker (UC Berkeley, 87), Shiferaw Beranu (Rutgers, 87), Martin Brown (Georgia Tech), Busiso Chisala (UC Berkeley), Curtis Clark (Michigan), Donald Ray Cole (Mississippi, 85), Kevin Corlette (Harvard), Melvin Currie(Pittsburgh, 83), Dennis Davenport (Howard), Arouna Davies (New Mexico State, 86), Nathaniel Dean (Vanderbilt, 87), George Edmunds (Old Dominion, 87), Dawit Getachew (Illinois Institute of Technology), James Ervin Glover (Auburn, 84), Lorenzo Hilliard (Maryland, 86), Abdulcadir Issa (Howard, 88), Matthew Kambule (Massachusetts), Shadrach Kwalar (Wayne State, 85), Amha Lisan (Howard, 88), Christopher Mawata (Hawaii at Manoa, 87), Walter Miller (CUNY Graduate School, 86), Godfrey Muganda (Leheigh), N'Ekwunife Muoneke (Houston, 85), Mark Muzere (Northwestern, 87), Ibula Ntantu (VPI), Levi Nyagura (Southern Illinois), Frank Odoom (Iowa State), Jan Persens (Cornell), De Juran Richardson (Northwestern, 87), Bonita Saunders, (Old Dominion, 85), Daphne Smith (MIT), Vernise Steadman (Howard, 88), Victoire Tankou (Southern Illinois), Hanson Umoh (Howard), Nathaniel Whitaker (UC Berkeley, 87), Leon Woodson (Howard), Paul Wright (UC Berkeley), Busa Xaba (Arizona), and Abdulkeni Zekeria (Howard, 84).

It is not the purpose of this paper to trace what has happened to these individuals since 1988. In passing, however, let me note that Ashford and Dean are on the NAM Board of Directors. Ashford, Clark, Miller, and Woodson presented research papers at this conference (Second Conference of African American Researchers in the Mathematical Sciences). Dean was one of the organizers of this conference and Woodson is one of the organizers for the next conference.

Those persons desiring more current information about these and other minority mathematicians may wish to contact the Mathematical Association of America. In February of 1995 the MAA received funding from the Sloan Foundation to establish an archival record of minority mathematicians with Ph.D.s and a directory of minority mathematicians. The grant was "to develop an Archival Record and Directory of Minority Mathematicians with Ph.D.s in Mathematics or Mathematics Education. Together they will contain the name, current address, and institutions from which the degrees were received with dates, ethnicity, and a brief biographical sketch of approximately 350 African American, Hispanic American, and American Indians. The Archival Record is historical in nature and the Directory will be a more contemporary document."

5. Feedback

Of particular interest to me was the correspondence I received from the department chairs. Many wrote "none," but several added comments. I want to thank the chairs who took the time to respond with comments like these:

- Wish we did. Please keep me posted.

- Unfortunately, no black student received a Ph.D. in mathematics from X in the past five years. We are very interested in recruiting minority students for our Ph.D. program.

• Our only black Ph.D. in mathematics during the past five years was X who received the degree this past spring. He is from Y and has returned to his country for employment. We would welcome any applications from...... Good luck on your NAM program for next January.

• I am sorry to report that we have had no Black recipients of the Ph.D. degree in the past five years. We hope to rectify this situation in the near future. To this end, we have recently constituted a departmental committee on Mathematics Education for Minority Students.

• X is regrettably the only Black mathematician to receive his doctorate from Y in the past five years. Y does have some excellent Fellowships for minority students. I would be interested in setting up a network of Black mathematicians to make known the opportunities at Y and our commitment to encourage a wide and diverse mathematical community.

• My records indicate there have been none. However, we do have two that have just passed their qualifying examinations.

• Unfortunately, we don't even have any in the pipeline (though we've tried)...Please send me details on NAM institutional memberships.

• Unfortunately, I must report that X has not graduated any Black with a Ph.D. in that period. There has been one M.Sc. awarded. At the present time we have one Black Ph.D. candidate and one master's candidate among the 185 graduate students in the department.

• I regret to report that no Black ..... We would be quite interested in your list. We have faculty openings and we are interested in increasing the number of Black mathematicians on our faculty (we currently have one).

• No black students have received Ph.D.'s from our Department in the past five years. But four black students are currently enrolled and we expect to have a total of five next year.... I would be interested in becoming a member of NAM.

• We have one black student, X, who should have his Ph.D. requirements completed by the end of this year. He is typing his dissertation now, and has a language requirement to complete. He will be our first.

• Your recent letter arrived just as we were thinking about recruiting qualified minority students for our graduate program.

• Although we have had five black Ph.D. students in the past, we have had none since 1981.

• We've had none (ever, I think). Certainly none in the past five years.

• Several years ago we had a number of Black Ph.D. candidates. Although we have graduated ten, none have graduated in the past five years.

These samples are fairly representative of the responses I received. Many expressed a sincere interest in recruiting minority students and seemed eager to receive suggestions from any source. In case you think that all were so pleasant, let me give you my favorite response to my memo to the presidents. Remember that I was doing all this correspondence from home with my $288 Amstrad computer, addressing the envelopes by hand, and licking stamps in my spare time.

• I am responding on behalf of X, President of Y, to the memo you sent on October 5, 1988, concerning...I am not at all pleased at the tenor and the wording of the letter you sent to President X. Your letter accuses the chairperson of our Department of Mathematics of not responding to two earlier messages from you. In fact, neither of the messages to which you refer was received by the Department of Mathematics. Even if they were, however, it strikes me as radically unprofessional of you to suggest that it was traits such as "indifference," "incompetence," or "racism" which caused the lack of response you claimed. I find such charges emanating from someone who didn't even take the time to learn the name of the president of our University, but addressed his memo simply to "The President," wholly unacceptable and entirely unprofessional...Your memo embarrasses you, Florida A&M University, and the National Association of Mathematicians. Sincerely, Z, Vice President for Academic Affairs.

That his reply had little relevancy to my memo seemed obvious, but I was still fuming about it to my NAM colleagues weeks later in Phoenix. Their response was "Welcome to the club. You are starting to understand what it means to be Black in America. We endure that kind of nonsense every day of our lives. Just concentrate on the good and forget the bad." That's not bad advice for anyone and soon I was happily busy with other NAM activities.

6. Final Remarks

Over the years I have been fortunate to use my mathematics background in endeavors on several continents in a variety of languages and cultures, but this project was rather special. It was a real joy to be involved in its development. I owe much to good friends from NAM like David Blackwell, J. Ernest Wilkens, Johnny Houston, Rogers Newman, Raymond Johnson, James Donaldson, Vivienne Mays, Harriett Walton, Eleanor Jones, and Robert and Sylvia Bozeman. We will not rest until African Americans take their rightful place within the mathematical community. Will you join us?

MATHEMATICS DEPARTMENT, FLORIDA A&M UNIVERSITY, TALLAHASSEE, FL 32307

DIMACS Series in Discrete Mathematics
and Theoretical Computer Science
Volume **34**, 1997

Dr. J. Ernest Wilkins, Jr.: The Man And His Works

(Mathematician, Physicist and Engineer)
A Biographical Preview*

NKECHI AGWU AND ASAMOAH NKWANTA

ABSTRACT. This paper is based on a research study conducted by the authors at the 1996 Mathematical Association of America (MAA) Institute in the History of Mathematics and Its Uses in Teaching (IHMT). The purpose of this paper is to highlight the significant contributions of the African-American educator and researcher in mathematics, physics and engineering - Dr. J. Ernest Wilkins, Jr.

1. INTRODUCTION

In order to acknowledge African-American mathematicians, scientists and engineers for their significant scientific and technical contributions, and to encourage many African-American students to major in scientific and technical fields, biographical literature of these African-American mathematicians, scientists and engineers is necessary. Dr. J. Ernest Wilkins, Jr. (see photograph) is an African-American hero because of his immense contributions to mathematics, physics, engineering and education. In particular, some of his contributions to nuclear science were instrumental in ending World War II. Another significant contribution of his has been, and still is, the grooming and mentoring of many African-Americans into science, engineering and mathematics related fields, at all levels of the scientific pipeline. His life story is an inspiration to the authors of this paper. Therefore, they have chosen to document his biography to acknowledge his contributions, and to foster pride within the African-American community.

---

* 1991 Mathematics Subject Classification. Primary 01A05.

(Photograph provided and permission for use given by Dr. J. Ernest Wilkins, Jr.)

2. EARLY CHILDHOOD

Dr. J. Ernest Wilkins, Jr. was born on November 27, 1923 in Chicago, Illinois. His mother was Lucile B. Wilkins (nee Robinson) and his father was J. Ernest Wilkins. Lucile Wilkins was an educator, and J. Ernest Wilkins (the father of Dr. J. Ernest Wilkins, Jr.) was an accomplished attorney, a presidential appointee as the assistant secretary of labor in 1954, a member of the Civil Rights Commission in 1958, and the first African-American to officially participate in a cabinet meeting (Burk, 1984).

Dr. Wilkins was an exceptional student throughout his childhood. He was a child prodigy. His reading skills developed very early. This early development enabled him to begin elementary school at the age of four. It also allowed him to skip several grade levels. While most children are in second grade at the age of seven, Dr. Wilkins was in fifth grade at this age.

Dr. Wilkins' strong interest in mathematics developed at an early age. This interest was encouraged by his parents. They always stressed to him the value of education and the importance of achievement. At Parker High School in Chicago, Illinois, he was fortunate to have a female teacher who recognized and encouraged his mathematical talent. She accelerated him from Algebra II to Algebra III. She got him enrolled in Geometry I and Geometry II, simultaneously. These positive home and school environments factored significantly in enabling him to graduate from Parker High School within a span of three years, at the tender age of thirteen.

3. HIGHER EDUCATION

Dr. Wilkins was immediately admitted to the University of Chicago, Chicago, Illinois, upon graduation from Parker High School. Still 13 years of age, he was the youngest student on campus. The university officials were concerned with his ability to do the required course work rather than his youthfulness. However, their concerns were allayed. The young Wilkins had the ability and motivation to take more than the required course work each quarter, and this is precisely what he did.

One of the reasons why Dr. Wilkins chose to attend the University of Chicago was that he had grown up hearing positive things about the university from his parents. His parents were alumni of this university. Another reason was the close proximity of the university to his home. The university was about one and one-half miles from his home, so he could easily commute between his home and the university.

While attending the University of Chicago as an undergraduate student, Dr. Wilkins majored in mathematics. He was also the University table tennis champion for three years, and he won the boys' states championship in 1938. Additionally, he participated in the William Lowell Putnam Mathematical Competition for undergraduates, and was ranked in the top ten in 1940. These factors facilitated his intellectual development. Other factors such as university scholarships for tuition, family support for other college expenses, and tutoring opportunities for pocket money were instrumental in enabling him to take more than the required course work each quarter, and to complete his undergraduate studies within three and one-half years.

Dr. Wilkins first gained nationwide attention when he graduated Phi Beta Kappa, at the age of 16, with a bachelors degree in mathematics from the University of Chicago

(*The Crisis*, September 1940). Interestingly, his father had also graduated Phi Beta Kappa with a bachelors degree in mathematics from the University of Illinois, 22 years earlier. At the time of his graduation, Dr. Wilkins had accumulated enough graduate credits to enable him to complete a masters degree in less than a year, and he used these credits in achieving this goal. In 1941, at the age of 17, he was the proud recipient of a masters degree in mathematics from the University of Chicago. Immediately afterwards, he continued on at this university to work on a doctoral degree in the same field, writing a dissertation in the area of calculus of variations.

The mathematical influence of Dr. Wilkins in the area of calculus of variations can be traced back to Felix Klein, Karl Weierstrass and other key mathematicians of the school of calculus of variations. Magnus R. Hestenes was the dissertation advisor of Dr. Wilkins. Dr. Wilkins chose Hestenes as his advisor because he had taken many courses with him and saw opportunities for possible research in the field of calculus of variations. Hestenes received his Ph.D. from the University of Chicago in 1932, working with G. Bliss as his dissertation advisor. Bliss received his Ph.D. from the University of Chicago in 1900. Oskar Bolza was Bliss' dissertation advisor. Bolza was a student of Felix Klein and Karl Weierstrass in Berlin, Germany, and Klein accepted his Ph.D. dissertation in 1886.

Dr. Wilkins completed his doctoral degree in 1942 at the age of eighteen. However, his Ph.D. was not conferred on him until he was 19 years old. He is one of the youngest persons, and the eighth African-American, to have received a Ph.D. in mathematics in the U.S. His doctoral dissertation, *Multiple Integral Problems in Parametric Form in the Calculus of Variations* (1942), was rewritten with the same title and published in the *Annals of Mathematics* (1944). He considers this dissertation study to be a minor part of his work because "it did not yield any ground breaking results."

Dr. Wilkins' formal education did not end with the doctorate in mathematics. Years later, he continued with his education by studying mechanical engineering at New York University (NYU), New York. He graduated from NYU, Magna cum Laude, with a bachelors in mechanical engineering in 1957, and with a masters in this field in 1960. Other noteworthy facts related to his higher education include the following: Sigma Xi (a scientific award) in 1942, Pi Tau Sigma (a mechanical engineering award) in 1956, and Tau Beta Pi (an engineering award) in 1956.

4. PROFESSIONAL EXPERIENCE

With the completion of his doctorate degree in mathematics, Dr. Wilkins embarked upon a professional journey that set him on course for his significant contributions to nuclear science and the education of African-American youth. He began as a postdoctoral research fellow at the Institute of Advanced Studies, Princeton, New Jersey, in 1942. Then, in 1943, he accepted a mathematics faculty position at Tuskegee Institute, Alabama, one of the Historical Black Colleges or Universities (HBCUs). He needed a job, and Tuskegee was the first institution to make him an offer. He left Tuskegee Institute in 1944 to work as a physicist in the metallurgical laboratory of the University of Chicago. This move initiated his professional journey into the world of nuclear physics, particularly in relation to atomic energy.

At the metallurgical lab, Dr. Wilkins primarily worked on the *Manhattan Project*. The Manhattan Project was the code name for the U.S. War Department's top secret plan to develop the atomic bomb of World War II that destroyed Hiroshima and Nagasaki in Japan, in 1945. However, Dr. Wilkins did not realize the goal of the Manhattan Project until August 7, 1945, a day after the bomb destroyed Hiroshima.

Dr. Wilkins also had the opportunity of working with several top scientists at the metallurgical lab. One of these scientists was Eugene Wigner, a Nobel Prize-winning Physicist, with whom he wrote the paper, *Effect of the Temperature of the Moderator on the Velocity Distribution of Neutrons With Numerical Calculations for H as a Moderator* (1944). Classified when written in 1944, this paper was declassified in 1948 and finally published in *Collected Works of Eugene Paul Wigner* (1992).

In 1946, despite warnings of institutional racism against African-Americans within American Optical Company in Buffalo, New York, Dr. Wilkins left the metalluri gical lab and went to work as a mathematician for this company. According to the September 1950 issue of *Ebony,* this was a period when many African-American scientists were being discriminated against in gaining access or upward mobility to decent jobs.

Dr. Wilkins left American Optical Company in 1950 to work as a senior mathematician at United Nuclear Corporation in White Plains, New York. He moved through the ranks from senior mathematician to become manager of the mathematics and physics department in 1958. It was while he was working as a mathematician for this organization that he was motivated to study mechanical engineering. Dr. Wilkins noticed that most of his engineering colleagues were trying to do the mathematics required for their projects without consulting the mathematicians until it was usually too late. This resulted in cost overruns which affected the productivity of the company. Dr. Wilkins surmised that if he were an engineer, his engineering colleagues might be willing to consult him at the beginning of their projects, thereby saving the company money and time. This was the motivating factor behind his decision to study mechanical engineering, which as indicated earlier he studied at New York University (NYU). It is a decision he considers to be one of the best decisions in his life.

At NYU one of Dr. Wilkins' engineering professors took an interest in mentoring him. His engineering colleagues at work learned through this mentor that he was studying to become *one of them*, and they began to consult him about engineering projects. It is noteworthy to point out that around this period in 1956 he was elected a fellow of the American Association for the Advancement of Science (AAAS).

In 1960, Dr. Wilkins moved on to work as an administrator for General Atomic Corporation in San Diego, California. He worked successfully through this position to become assistant chairman of the theoretical physics department, and then assistant director of the John Jay Hopkins Laboratory in 1965. He later became director of the defense science and engineering center and finally director of computational research. It was during this period that he was a recipient of the following honors: a visiting lectureship of the Mathematical Association of America (MAA) from 1963 to 1967, and a fellow of the American Nuclear Society (ANS) in 1964.

A few years later, Dr. Wilkins was approached by Dr. Warren Henry of the physics department at Howard University, Washington D.C., about a position within this department. He was later invited to visit the campus and an offer was made. The Howard University position was welcomed by Dr. Wilkins, because it came at a point when he wanted a career change. In 1970, he moved from industry back to academia when he accepted the position of distinguished professor of applied mathematical physics at Howard University, another HBCU.

At Howard University, Dr. Wilkins supervised four doctoral dissertations: *Apodization With Specified Transmitance* (1974) by Cleo L. Bentley, *Optimal Optical Systems: Two Concentric Spherical Mirrors* (1975) by Winston Thompson, *Minimum*

Critical Mass Nuclear Reactors (1977) by Keshava N. Srivastava, and *Apodization of Optical Systems With Poor Resolving Power* (1979) by John N. Kibe. John Kibe's dissertation was supervised to completion by Dr. Wilkins, even though he left Howard University in 1977.

During the period Dr. Wilkins was a faculty member at Howard University, he also served as an advisor in 1975 for establishing the doctoral degree program in mathematics. From 1976 to 1977, Dr. Wilkins worked as a visiting scientist at Argonne National Laboratory in Argonne, Illinois, while on sabbatical leave from Howard University. In 1976, he was inducted into the National Academy of Engineering (NAE). The NAE is a private organization established in 1964 under the congressional character of the National Academy of Sciences (NAS) to advise the federal government on questions of science and technology. Membership in the NAE is based on the significant contributions that a person has made to engineering theory and practice. The citation of Dr. Wilkins' induction into the NAE reads: "Peaceful application of atomic energy through contributions to the design and development of nuclear reactors."[1] A major contribution of Dr. Wilkins to engineering theory is the development of mathematical models by which the amount of gamma radiation absorbed by a given material can be calculated (National Association of Mathematics (NAM), 1988).

Dr. Wilkins returned to industry as vice president and associate general manager for science and engineering of EG&G Idaho Incorporated in Idaho Falls, Idaho, after leaving Howard University in 1977. In 1978, he became deputy general manager for science and engineering for this company, while maintaining his vice president status. In 1984, he retired from EG&G. While on retirement, he was honored as a fellow from 1984 to 1985 at the Argonne National Laboratory in Argonne, Illinois.

Dr. Wilkins continued in retirement until 1990 when he accepted a distinguished professorship of applied mathematics and mathematical physics at Clark Atlanta University (CAU), another HBCU. Similar to the case of Howard University, he never formally applied for the position. He was approached by the then and current president and invited to visit the university. An offer was made that he considered worthy enough to bring him out of retirement. To date, he is still a distinguished professor of applied mathematics and mathematical physics at CAU.

---

1 NAE archival data.

One of the motivating factors for Dr. Wilkins' acceptance of the CAU position was Albert Bharucha-Reid, an African-American male mathematician at this university. Bharucha-Reid was actively involved in, and a forerunner of, research in the area of random polynomials. This is Dr. Wilkins' present area of research. He was hoping to work in collaboration with Bharucha-Reid. However, by the time he started working at CAU Bharucha -Reid was deceased.

Bharucha-Reid's contribution to mathematics continues through the work of Dr. Wilkins. The following are the titles of a few publications by Dr. Wilkins in the area of real zeros of a random polynomial: *An Upperbound for the Expected Number of Real Zeros of a Random Polynomial* (1973), *An Asymptotic Expansion for the Expected Number of Real Zeros of a Random Polynomial* (1988), *Mean Number of Real Zeros of a Random Trigonometric Polynomial* (1991), *Mean Number of Real Zeros of a Random Trigonometric Polynomial II* (1995), and *Mean Number of Real Zeros of a Random Trigonometric Polynomial III* (1995).

It is also noteworthy to point out that Dr. Wilkins was president from 1974 to 1975 of the American Nuclear Society (ANS), and that he was a recipient of the following awards and honors: the U.S. Army Outstanding Civilian Service Medal (1980), the Quality Engineering for Minorities (QEM) Network Giant in Science Award (1994), the National Association of Mathematics (NAM) Lifetime Achievement Award (1994), and the U.S. Department of Energy Special Recognition Award (1996). The inscription on the plaque presented by NAM reads:

> *In recognition of more than fifty professional years as a world class mathematician, physicist and engineer, gifted teacher and productive scholar it is NAM's distinct pleasure to name in your honor,*
> *THE J. ERNEST WILKINS, JR. LECTURE*
> *To be given annually at the NAM undergraduate MATHFest conference and proclaiming your lecture today,*
> *"Real Zeros of Random Polynomials,"*
> *to be the inaugural J. Ernest Wilkins, Jr. lecture.*[2]

5. CONCLUSION

Dr. Wilkins has numerous published and unpublished works in many areas of mathematics, science and engineering, viz, differential and integral equations, projective

---

2 NAM Newsletter.

differential geometry, special function theory, calculus of variations, optics, heat transfer and nuclear transport theory. To date, he has 84 research publications, two additional research papers accepted for publication and 21 unpublished, unclassified, atomic energy reports. Dr. Wilkins has been a consultant for numerous companies and academic institutions, and an active participant in many professional and civic organizations, including AAAS, American Mathematical Society (AMS), ANS, American Society for Mechanical Engineers, MAA, NAE, NAM, Optical Society of America and the Society for Industrial and Applied Mathematics (SIAM). The importance of his scientific contributions is evidenced by these facts and the several honors that have been bestowed upon him. Ironically, his honors do not include any honorary degree to date.

Interestingly, Dr. Wilkins' professional experience as a full-time faculty member in academia has been at HBCUs. This has provided African-American students at these colleges or universities close contact with an African-American mathematician, physicist and engineer who has made significant contributions to the advancement of these fields. This affiliation has enabled Dr. Wilkins to groom and mentor numerous African-Americans into science, engineering and mathematics related careers.

This paper is a preview of the biographical work in progress of Dr. Wilkins by the authors. Hopefully, it will inspire others to write biographies that document the accomplishments of great African-American mathematicians, scientists and engineers. The works by Albers and Alexanderson (1985), Donaldson (1989), Hattie (1977), Kenschaft (1981, 1987), Low and Clift (1981), Newell (1980) and Sammons (1990) are a few trailblazing examples that helped to inspire the authors to write this paper.

ACKNOWLEDGMENTS. The authors would like to thank the following people for their contributions to this article:

(i) Dr. J. Ernest Wilkins, Jr. for allowing the authors to interview him, for providing the authors with documented facts about himself, for giving the authors permission to document this biography and use his photograph, and for editorial contributions.

(ii) Dr. Warren Henry for allowing the authors an interview about his knowledge of the life and works of Dr. Wilkins, and for providing the authors with copies of public documents relating to Dr. Wilkins from the archives of the Department of Physics and Astronomy at Howard University.

(iii) Dr. James Donaldson for providing the authors with biographical notes related to the life and works of Dr. Wilkins.

(iv) The MAA and NAE for providing the authors with archival data related to the Dr. Wilkins.

(v) Dr. Charles Pierre for providing the authors with Dr. Wilkins' telephone number, and copies of public documents relating to Dr. Wilkins from the archives of Clark Atlanta University.

(vi) Drs. Victor Katz, Frederick Rickey and Steven Schot, the organizers of the Institute in the History of Mathematics and Its Uses in Teaching (IHMT), for providing the authors with the initial opportunity to present Dr. Wilkins' biography in a seminar
(vii) Dr. Auset BaKhufu for her editorial contributions.
(viii) Dr. Geoffrey Akst for his editorial contributions.

BIBLIOGRAPHY

Agwu, N. and Nkwanta, A., *Transcript From an Interview With Dr. J. Ernest Wilkins, Jr.*, (1996).

_____, *Transcript From an Interview With Dr. Warren Henry*, (1996).

Albers, D.J. and Alexanderson G. L (eds.), *Mathematical People*, Boston, MA: Birhauser, (1985).

American Mathematical Society (AMS), *Monthly*, vol. 47, (1940), 331.

Bentley, C. L., *Apodization with Specified Transmitance* (Dissertation), Washington D.C.: Howard University, (1974).

Bliss, G. A., *Oskar Bolza: In Memorium*, *Bulletin of the American Mathematical Society*, vol. 50, (1944), 478-489.

Burk, R. F., *The Eisenhower Administration and Black Civil Rights,* Knoxville, TN: University of Tennessee Press, (1984).

Donaldson, J., *Biographical Notes Related to the Life and Works of Dr. Wilkins* (1989).

_____, *Black Americans in Mathematics*, *A Century of Mathematics in America, Part III*, Providence, RI: AMS, (1989), 449-469.

Hattie, C., *Blacks in Science: Astrophysicist to Zoologist*, Hicksville, NY: Exposition Press, (1977).

Kenschaft, P.C., *Black Women in Mathematics in the United States*, *American Mathematical Monthly*, vol. 88, (1981), 592-604.

_____, *Black Men and Women in Mathematical Research*, *Journal of Black Studies*, vol. 18, (1987), 170-190.

Kibe, J. N., *Apodization of Optical Systems With Poor Resolving Power* (Dissertation), Washington D.C.: Howard University, (1979).

Low A. and Clift V. A. (eds.), *Encyclopedia of Black America*, New York, NY: McGraw-Hill, (1981).

"Many Scientist Still Find Bias Bars Way to Decent jobs," *Ebony*, (September 1950), 20.

Newell, V. K. (ed.), *Black Mathematicians and Their Works*, Ardmore , PA: Dorrance, (1980).

National Academy of Engineering (NAE), *Public Directory of Members and Foreign Associates*, Washington D.C.: NAE, (July 1995).

National Association of Mathematics (NAM), *Newsletter*, Washington D.C.: NAM, (Winter 1988).

Parshall, K. H. and Rowe, D. E., *Emergence of the American Mathematical Research Community, 1876 - 1900: J. J.. Sylvester, Felix Klein, and E.H. Moore*, Providence, RI: AMS, (1994).

"Phi Beta Kappa at 16," *The Crisis*, (September 1940), 288.

Sammons, V. O., *Blacks in Science and Medicine*, New York, NY: Hemisphere Publishing Corporation, (1990).

Srivastava, K. N., *Minimum Critical Mass Nuclear Reactors* (Dissertation), Washington D.C.: Howard University, (1977).

Thompson, W., *Optimal Optical Systems* (Dissertation), Washington D.C.: Howard University, (1975).

Wilkins, J. E., Jr., "Multiple Integral Problems in Parametric Form" (Published Dissertation), *Annals of Mathematics*, vol. 45, (1944), 312-334.

_____, "An Upperbound for the Expected Number of Real Zeros of a Random Polynomial," *Journal of Mathematical Analysis and Applications*, vol. 42, (1973), 569-577.

_____, "An Asymptotic Expansion for the Expected Number of Real Zeros of a Random Polynomial," *Proceeding of the American Mathematical Society*, vol. 103 (4), (1988), 1249-1258.

_____, "Mean Number of Real Zeros of a Random Trigonometric Polynomial," *Proceedings of the American Mathematical Society*, vol. 111 (3), (1991), 851-863.

_____, "Mean Number of Real Zeros of a Random Trigonometric Polynomial II," *Topics in Polynomials of One and Several Variables and Their Applications*, Singapore.: World Scientific Publishing Company, (1993), 581-594.

Wilkins J. E., Jr. and Souter S. A., "Mean Number of Real Zeros of a Random Trigonometric Polynomial III," *Journal of Applied Mathematics and Stochastic Analysis*, vol. 8, (1995), 299-317.

Wilkins J.E., Jr. and Wigner, E.P., "Effect of the Temperature of the Moderator on the Velocity Distribution of Neutrons With Numerical Calculations for H as a Moderator," in *Collected Works of Eugene Paul Wigner, Part A, Volume V, edited by A. M. Weinberg,* Berlin, Germany: Springer Verlag, (1992), 499-508.

DEPARTMENT OF MATHEMATICS, BOROUGH OF MANHATTAN COMMUNITY COLLEGE (BMCC), CITY UNIVERSITY OF NEW YORK (CUNY), NEW YORK, NY, 10007

E-mail address: nmabm@cunyvm.cuny.edu

DEPARTMENT OF MATHEMATICS, HOWARD UNIVERSITY, WASHINGTON, D.C. 20059

E-mail address: nkwanta@scs.howard.edu

Selected Titles in This Series

(Continued from the front of this publication)